John Howard Appleton

Chemistry

Developed by Facts and Principles Drawn Chiefly from the Non-Metals

John Howard Appleton

Chemistry

Developed by Facts and Principles Drawn Chiefly from the Non-Metals

ISBN/EAN: 9783744749886

Printed in Europe, USA, Canada, Australia, Japan

Cover: Foto ©berggeist007 / pixelio.de

More available books at **www.hansebooks.com**

A Chemist's Work Bench.

CHEMISTRY:

DEVELOPED BY

FACTS AND PRINCIPLES DRAWN CHIEFLY FROM THE NON-METALS,

BY

JOHN HOWARD APPLETON, A. M.,

Professor of Chemistry in Brown University.

AUTHOR OF

"THE YOUNG CHEMIST,"
"QUALITATIVE CHEMICAL ANALYSIS,"
"QUANTITATIVE CHEMICAL ANALYSIS,"
"THE LABORATORY HANDBOOK."

PROVIDENCE:
PROVIDENCE LITHOGRAPH COMPANY.
1884.

OTHER WORKS ON CHEMISTRY

BY

Professor Appleton:

I. The Young Chemist: A book of chemical experiments for beginners in Chemistry. It is composed almost entirely of experiments, those being chosen that may be performed with very simple apparatus.

II. Qualitative Analysis: A brief but thorough manual for laboratory use.

It gives full explanations and many chemical equations. The processes of analysis are clearly stated and the whole subject is handled in a manner that has been highly commended.

III. Quantitative Analysis: The treatment of the subject is such as to afford an acquaintance with the best methods of determining all the principal elements, as well as with the most important type-processes both of gravimetric and volumetric analysis.

THE EXPLANATIONS ARE DIRECT AND CLEAR so that a pupil is enabled to work intelligently *even without the constant guidance of the teacher*. By this means the book is adapted for self-instruction of teachers and others who require this kind of help to enable them to advance beyond their present attainments.

IV. The Laboratory Handbook: An annual publication containing many convenient tables for laboratory use. New tables are constantly introduced, and changes are made in order to keep the matter abreast of the latest discoveries.

PREFACE.

This little book has been prepared as a popular introduction to the study of chemistry.

It is probably needless to recommend the *subject*: chemistry is recognized as a science of such general interest, such wide usefulness, and such universal application, that no intelligent person can endure long to remain ignorant of its principal facts and laws.

This book treats principally of the *non-metals;* it is believed to be the verdict of authors and teachers of experience, that these furnish the most suitable material for a beginner in the study of chemistry; these best present the fundamental facts and principles of the science, and they do it in connection with objects and phenomena easily accessible to almost every civilized human being.

The author contemplates the preparation hereafter of a book of similar general character, only having its principal facts drawn from the chemistry of the *metals*.

In writing this book it has been the effort to treat the subject in a style that shall be attractive to the general reader; but it is believed that in no case has scientific fact been sacrificed in the interest of popular form.

The arrangement of matter in the book is in accordance with the following plan: After the introductory chapters, which present the general principles of chemical action, the chief non-metals are treated in a scientific order as follows: the *monads*, hydrogen, chlorine, bromine, iodine, and fluorine; then the *dyads*, oxygen and sulphur; next the *triads*, boron nitrogen; finally the *tetrads*, carbon and silicon · thus including the four great groups into which the non-metals are naturally arranged.

The *historical and biographical sketches*, that are distributed through the book, have been introduced with the view of legitimately helping to retain the reader's attention. Most of the *experiments* described are such as may be performed by any one possessed of reasonable skill; it is believed that they will afford profitable instruction as well as entertainment. Allusions to the *applications of chemistry* to the affairs of everyday life have been carefully introduced and have been developed as fully as the circumstances seem to warrant.

Perhaps some comment upon the *reading references* (pp. 10, 17, 24, 27, 31, 38, 45, 56, 73, 84, 98, 111, 116, 133, 156, 161, 186, 196, 214,) is proper. They are largely from periodical publications, and it is thought that they will be of service, especially to mature students. Helps to reading are now viewed as among the most important offerings of teachers to learners. The reading lists in this book point out some papers which are selected as being chiefly popular in style; if these lead the reader to consult the others he will find himself introduced to some of the most important contributions to the knowledge of our science.

BROWN UNIVERSITY, August, 1884.

CONTENTS.

CHAPTER I.—The scope of chemistry, ... 7
CHAPTER II.—The elementary substances, ... 12
CHAPTER III.—Names and symbols of elements, . . . 19
CHAPTER IV.—Classification of the elementary substances; metals and non-metals, ... 28
CHAPTER V.—Compound substances; examples of binary and ternary compounds, ... 32
CHAPTER VI.—The construction of substances; the mass, the molecule, the atom, ... 39
CHAPTER VII.—How chemical affinity works; the modern atomic theory, ... 46
CHAPTER VIII.—Hydrogen; its discovery; method of preparation; its properties, ... 58
CHAPTER IX.—Balloons; their invention and uses, . . . 74
CHAPTER X.—Chlorine; its preparation and properties; chlorohydric acid; bleaching powder, ... 85
CHAPTER XI.—Bromine, ... 100
CHAPTER XII.—Iodine, ... 105
CHAPTER XIII.—Fluorine, ... 112
CHAPTER XIV.—Oxygen; its discovery; its preparation; its properties; its relation to hydrogen; the compound blowpipe; its relation to combustion in general, and to animal respiration, ... 117

CHAPTER XV.—Water, its importance to living beings; its terrestrial circulation; its influence on climate, . . . 134
CHAPTER XVI.—Sulphur; sulphuretted hydrogen; sulphur dioxide, ... 142
CHAPTER XVII.—Sulphur trioxide; manufacture of sulphuric acid, ... 151
CHAPTER XVIII.—Boron, ... 157
CHAPTER XIX.—Nitrogen; its discovery; its properties; compound with hydrogen; compounds with oxygen; nitric acid, ... 162
CHAPTER XX.—The Atmosphere, ... 170
CHAPTER XXI.—Explosives; gunpowder; fireworks; fulminates gun-cotton; nitroglycerine, ... 177
CHAPTER XXII.—Phosphorus; friction matches, . . . 187
CHAPTER XXIII.—Carbon; charcoal; lampblack; coal; graphite; the diamond; other natural forms of carbon, - - - 197
CHAPTER XXIV.—Compounds of carbon and oxygen, . . 215
CHAPTER XXV.—Illuminating gas. 221
CHAPTER XXVI.—Silicon, ... 228

CHEMISTRY

CHAPTER I.

THE SCOPE OF CHEMISTRY.

CHEMISTRY treats of all kinds of material substances. This is a very broad assertion, but it appears to be true. The solid rock matter of the earth and the wealth of living animal and vegetable substances upon it undergo all their varied changes in subjection to chemical laws. The same is true of the water and of all liquid things we know; and the declaration applies yet further to the invisible gaseous mass which surrounds and envelopes our terrestrial globe. This deeper but thinner ocean which we call the atmosphere is also governed by chemical law in all its varied relations to the living beings as well as to the inanimate substances that have their existence within it. In thought we may ascend above these solid liquid and gaseous substances connected with our earth. When thus we reach out to the heavenly bodies beyond, we feel sure that these, possessing as they may, solid, liquid or gaseous

matter, are likewise controlled by chemical laws, and that in their changes they exemplify with more or less fulness, distinct chemical principles.

The Great Number of Different Substances in the Earth.

Here then it is intimated that chemistry relates to an enormous number of substances. In fact the number of the various kinds of matter already existing on our earth is so great that they have never been so much as counted, much less described, in any list or volume; nay more, doubtless many exist that human beings have never recognized at all. This last statement refers not merely to such substances as may be known only to savages dwelling beyond the reach of civilization and commerce, nor yet to such as may be secreted in absolutely uninhabited portions of the globe, nor even to those that exist so deep in the earth that man's power may never be sufficient to reach them. Probably even some of the most humble and familiar natural things, such as blades of wheat, petals of daisies, silks of corn, and the like, contain small quantities of distinct and separate compounds that have not yet been recognized as such by even the most skillful chemists.

But even compounds such as are described in books on chemistry are exceedingly numerous; and further, the chemical laws now known suggest the possibility of producing *artificially* a great multitude of substances not yet recognized, and even more than have yet been produced in the great laboratory of nature.

How it Happens that there is such a Variety.

By searching aright for the secret of the countless number and the rich and splendid variety of beings that nature and art present to the curious gaze of man, a comprehensive answer is at last obtained.

Forms of ordinary matter may be compared to great cathedrals, like those of Cologne and of Milan, which have been growing for centuries and which by the combined labor of artists and artisans have at length become intricate and beautiful structures, the admiration and delight of the beholder. Just as these arise from the combination in a multitude of ways of a comparatively small number of original and fundamental substances—like stone, brick, iron, copper, plaster, glass, wood—so all things known to chemists are made up of a few simple substances, either existing alone or in richly various combination.

The simplest substances when alone are called the *chemical elements* or elementary substances; the things resulting when different elements are united together are called *compounds*. Thus metallic iron is one familiar example of a chemical element; the oxygen gas of the atmosphere is another example. A piece of iron exposed to damp air soon becomes changed to a mass of iron rust. This rust is a compound; it is made up of iron and oxygen united together.

In the light of what has been said the chemical elements assume a new and grand importance: they are the individuals chosen by the Creator to be the foundation stones and the essential constituents of the glorious natural edifices of his handiwork.

Again when the elementary individuals unite together they do so by reason of the interaction of many and complex forces which reside, almost like soul and spirit, within the elements.

These last remarks suggest the two-fold character of chemical study. It involves, *First*, the examination of elementary substances and their compounds. *Secondly*, it requires a consideration of the many general and special *laws and forces* which determine the various possible combinations.

READING REFERENCES.

In general, the first-mentioned books in each group are those which are most accessible and at the same time most serviceable. Some rare and costly books are mentioned, for the benefit of persons who have access to large libraries.

Chemistry, General and Applied, Serial Publications.
    Chemical News. (William Crookes, Ed.) London. Weekly. (Commenced 1860.)
    Popular Science News and Boston **Journal of Chemistry.** Boston. Monthly. (Commenced 1867.)
    Journal of the Chemical **Society.** London. Monthly.
    ——Index to foregoing. 1841-1872: pp. 263.
    Annales de Chimie et de Physique. Paris. Monthly.
    ——**Table des Tomes I à XXX.** (1841-1851.) Paris. pp. 134.
    ——**Table Analytique des Tomes XXXI à LXIX.** 3d Series. (1851-1863.) Paris. pp. 474.
    ——**Table des Noms d'Auteurs et Table Analytique des Matieres.** (1864-1873.) 4th Series. Paris. pp. 249.
    Berichte der Deutschen Chemischen Gesellschaft. Berlin. (Commenced 1868.) 20 parts per year.
    Wagner, Johannes R. v.—Jahres-Bericht über die Fortschritte und Leistungen der chemischen **Technologie.** Leipzig. Annual. (Commenced 1855; last vol. had 1,332 pp.)
    ——Index to foregoing. Vols. I-X.
    ——   "   "   Vols. X-XX.

Dictionaries of Chemistry, etc.
    Watts, Henry.—Dictionary of Chemistry and the allied branches of other sciences. 8 vols. London. 1865-1875.
    Storer, Frank H.—First Outlines of a Dictionary of Solubilities of Chemical Substances. Cambridge. 1864.
    Wurtz, Ad.—Dictionnaire de Chimie, pure et appliquée. 3 vols. Paris. 1870.
    Fehling, Hermann v.—Neues Handwörterbuch der Chimie. A to Morphin. Braunschweig. 1871—now issuing.

General Treatises on Chemistry.

Roscoe, H. E., and Schorlemmer, C.—A Treatise on **Chemistry.** **London and New York.** 1878. **Vol. I, pp.** 771; Vol. II, part I, pp. 504, part II, **pp. 552;** Vol. III, part I, pp 724. (Now issuing.)

Cooke, Josiah P., Jr.—Principles of Chemical Philosophy. Boston. 1881.

Gmelin, Leopold (Henry Watts, Tr.)—Hand-Book of Chemistry. Printed for the Cavendish Society. 14 vols. London. **1848–1860.**

Graham-Otto's Ausführliches Lehrbuch der Chemie. **6 vols. Braunschweig.** 1857.

Schützenberger, P.—Traité de Chimie générale. **3 vols. Paris. 1880.**

CHAPTER II.

THE ELEMENTARY SUBSTANCES.

N the following page is a list of the sixty-six elementary substances now generally recognized as such. Their respective symbols and their atomic weights, both in exact and approximate numbers, are also given.

These substances, then, nearly seventy in number, are those from which are made up all material things now known to man. While it is not necessary for any one to retain such a list in memory, every person who desires any considerable knowledge of chemistry should be acquainted with each name and the symbol attached to it, and should know something of the natural sources and the properties of the substances designated.

Six Suggestions Conveyed by this Table.

A careful and intelligent reading of the list affords several important suggestions. The following are some of them:—

First. *The elements are not very numerous.* They are in fact very few, as compared with the countless number of substances they may form by their proper combinations.

Second. *They are however sufficiently numerous to produce the many substances recognized in nature.* For, consider how human language may have many words and yet all these may be spelled out by

The Chemist's Elementary Substances.

Name of Element.	Atomic Symbol.	Exact Atomic Weight.	Approximate Atomic Weight.	Name of Element.	Atomic Symbol.	Exact Atomic Weight.	Approximate Atomic Weight.
Aluminum..	Al..................	27.0090	27	Molybdenum	Mo..............	95.5270	95.5
Antimony...	Sb (Stibium)......	119.9550	120.	Nickel......	Ni.................	57.9280	57.9
Arsenic.....	As..................	74.9180	74.9	Niobium...	Nb.................	93.8120	93.8
Barium. ...	Ba	136.7630	136.8	Nitrogen....	N..................	14.0210	14.
Bismuth....	Bi.................	207.5230	207.5	Osmium	Os................	198.4940	198.5
Boron......	B..................	10.9410	10.9	Oxygen.....	O..................	15.9633	16.
Bromine ...	Br.................	79.7680	79.8	Palladium...	Pd.................	105.7370	105.7
Cadmium...	Cd.................	111.8350	111.8	Phosphorus.	P..................	30.9580	31.
Caesium...	Cs.................	132.5830	132.6	Platinum....	Pt.................	194.4150	194.4
Calcium....	Ca.................	39.9900	40.	Potassium ..	K (Kalium)......	39.0190	39.
Carbon.....	C...................	11.9736	12.	Rhodium....	Rh................	104.0550	104.1
Cerium.....	Ce.................	140.4240	140.4	Rubidium...	Rb................	85.2510	85.3
Chlorine....	Cl.................	35.3700	35.4	Ruthenium..	Ru................	104.2170	104.2
Chromium..	Cr.................	52.0090	52.	Scandium...	Sc................	43.9800	44.
Cobalt......	Co.................	58.8870	58.9	Selenium...	Se................	78.7970	78.8
Copper.....	Cu (Cuprum).....	63.1730	63.2	Silicon......	Si................	28.1950	28.2
Didymium..	D..................	144.5730	144.6	Silver	Ag (Argentum)...	107.6750	107.7
Erbium.....	E..................	165.8910	165.9	Sodium.....	Na (Natrium)....	22.9980	23.
Fluorine....	Fl..................	18.9840	19.	Strontium...	Sr.................	87.3740	87.4
Gallium.....	Ga.................	68.8540	68.9	Sulphur.....	S..................	31.9840	32.
Glucinum...	G or Be (Beryllium)	9.0850	9.1	Tantalum...	Ta.................	182.1440	182.1
Gold........	Au (Aurum).......	196.1550	196.2	Tellurium...	Te.................	127.9600	128.
Hydrogen...	H..................	1.0000	1.	Thallium....	Tl.................	203.7150	2 3.7
Indium.....	In.................	113.3980	113.4	Thorium ...	Th................	233.4140	233.4
Iodine......	I...................	126.5570	126.6	Tin..........	Sn (Stannum)....	117.6980	117.7
Iridium.....	Ir..................	192.6510	192.7	Titanium....	Ti.................	47.9997	48.
Iron........	Fe (Ferrum)......	55.9130	55.9	Tungsten...	W (Wolframium)..	183.6100	183.6
Lanthanum.	La.................	138.5260	138.5	Uranium....	U..................	238.4820	238.5
Lead........	Pb (Plumbum)....	206.4710	206.5	Vanadium ..	Va................	51.2560	51.3
Lithium.....	Li.................	7.0073	7.	Ytterbium..	Yb................	172.7610	172.8
Magnesium.	Mg.................	23.9590	24.	Yttrium.....	Y..................	89.8160	89.8
Manganese.	Mn.................	53.9060	53.9	Zinc	Zn................	64.9045	64.9
Mercury ...	Hg (Hydrargyrum)	199.7120	199.7	Zirconium..	Zr................	89.3670	80.4

combinations of few letters. Some English dictionaries register over a hundred thousand words, yet these are all made by the combinations of less than thirty letters. Now it is easy to comprehend how the few letters of an alphabet may be even still further combined in various ways so as to produce additional words almost without limit : in a similar manner it may be easily imagined that the sixty-five elementary substances have ample capabilities for giving rise not only to the compounds now known, but to yet more and more, almost without limit. It is true that the chemist discovers that some of the chemical elements appear to have a limited power of union, but in others he finds an apparently unbounded capacity to form new arrangements and combinations.

Third. The elements are mostly uncommon. Only about one-sixth of them possess names that are familiar to ordinary readers. Thus carbon, copper, gold, iron, lead, mercury, nickel, silver, sulphur, tin, zinc, are almost the only ones in the list that can be said to suggest familiar things. Indeed some members of this list exist in the earth in extremely small quantities; but man by his ingenuity and industry has gathered up even these and brought them near to the hand of every civilized being. Thus gold exists in the earth—so far as man has access to the earth—in only very minute amounts; yet gold has a multitude of common uses beside its employment in coinage. Various forms of decorative art, like gilded lettering on books, afford familiar examples. So also mercury, which in the ordinary thermometer is very familiar to every one, exists in the earth in but minute amounts.

When the chemist examines still more narrowly the composition of the terrestrial globe, he discovers an inequality yet more extraordinary than that hinted at. Thus it appears that probably one-half of our entire planet consists of a single substance (that is, oxygen) and that one-quarter of it consists of another single substance (that is, silicon). Since an amount equal to three-quarters of the earth's matter, by weight, is made up of but two elements, the remaining ones must necessarily exist in much smaller proportions.

The following table, given by Roscoe and Schorlemmer, shows the average composition of the earth's crust—so far as it is accessible to human investigation by means at present known:

Percentage Composition of the Earth's Solid Crust
(by Weight).

Oxygen,	44.0	to	48.7	per cent.
Silicon,	22.8		36.2	"
Aluminum,	9.9		6.1	"
Iron,	9.9		2.4	"
Calcium,	6.6		0.9	"
Magnesium,	2.7		0.1	"
Sodium,	2.4		2.5	"
Potassium,	1.7		3.1	"
	100.0		100.0	

According to this table, the sum total of the amounts of *all the elements not mentioned* may be estimated as less than one-tenth of one per cent. of the whole. This statement is rendered all the more striking when it is considered that in this minute fractional part must be included all coal and all the useful metals, except iron.

Another authority[*] declares that it is probable that an amount equal to ninety-nine one-hundredths of the entire weight of the solid, liquid and gaseous matter of our globe, is made up of only thirteen elementary substances. The elements referred to and their relative proportions are approximately represented in the diagram following:

---

[*]Professor J. P. Cooke.

Diagram of the Composition of our Globe (by Weight.)

SILICON, $\frac{1}{4}$	SULPHUR, HYDROGEN, CHLORINE, NITROGEN,	53 OTHERS
	POTASSIUM, SODIUM, IRON, CARBON,	
	ALUMINUM, MAGNESIUM, CALCIUM, $\Big\} \frac{1}{6}$	
OXYGEN, $\frac{1}{2}$		

Fourth. Most of the elements are metals. This may not appear to the ordinary reader until he is informed that terminations in *um*, as in case of aluminum, barium, cadmium, calcium and others are intended to suggest that the substances so designated are metals. Most of the other elements having names not terminating in *um* are called non-metals.

Fifth. Each chemical element has an atomic symbol, an abridgement, in some form, of its name.

Sixth. Each chemical element has an atomic weight. As the atomic weight of hydrogen is 1, without any fraction, it is easily understood that the weight of one atom of hydrogen is taken as the unit of the system. An inspection of the numbers given shows that in many cases the atoms weigh amounts that are very nearly exact multiples of the weight of an atom of hydrogen.

READING REFERENCES.

Atomic Weights, Calculations of

Becker, George F.—Atomic Weight Determinations: a digest of the investigations published since 1814. (Published as Part IV of the *Constants of Nature*, in Smithsonian Miscellaneous Collections, No. 358.) 1880.

Clarke, Frank W.—A Recalculation of the Atomic Weights. (Published as Part V of the *Constants of Nature*, in Smithsonian Miscellaneous Collections, No. 441.) 1882.

——Am. Chem. Jour. iii, 263. (1881.)

Atomic Weights, Periodicity of

Meyer, Lothar.—Chem. News. xli, 203.

Atomic Weights, Mendelejeff's Law of

Am. Chem. Jour.—iii, 455.
Cooke, J. P.—Chem. Philosophy. p. 265.
Wurtz, Ad.—Atomic Theory. p. 154.

Atomic Weights, Arithmetical Relations of

Hodges, M. D. C.—Silliman's Journal, 3d Ser. x, 277.
Newlands, J. A. R.—Chem. News. xlix, 198.

Atomic Weight of Oxygen.

Odling, W.—Jour. of Chem. Soc. of London. xi, 107.

Atomic Weight of Thallium.

Cookes, Wm.—Chem. News. xxix, 14, 29, 39, 55, 65, 75, 85, 97, 105, 115, 126, 137, 147, 157.

Atomic Weights, Prout's Hypothesis of

Cooke, J. P.—Chemical Philosophy, 270.
Clarke, F. W.—Am. Chem. Journal, iii, 272. (1881.)
Gerber,—Silliman's Journal, 3d Ser. xxvi, 236.

Atoms, Absolute Weight of

Annaheim, J.—Jour. of Chem. Soc. of London, xxxi, 31.

Elements, Defunct

Bolton, H. C.—American Chemist, i, 1.

Elements, Suggestions that Elements are Compound.

Lockyer, J. N.—Nature, Jan'y 2 and 9, also Nov. 6, 1879.
Hastings, C. S.—Criticism of above. Am. Chem. Jour., i, 15.

CHAPTER III.

NAMES AND SYMBOLS OF ELEMENTS.

NY history of the chemical elements, distinctly points to the enormous stride which chemical discovery has taken within the last hundred years. The dawn of this period was marked by many most important results: among these may be mentioned the detection of the elementary gases, oxygen, hydrogen and nitrogen. The light which these great events threw upon the future of the science enabled the chemists of that early period to perceive that the number of new compound substances then discovered, and likely soon to be discovered, called for a multitude of new terms. In 1787 the eminent French chemist, Lavoisier, in committee with Guyton de Morveau and others of their chemical associates of the French Academy, suggested a system by which a considerable number of chemical compounds, both then known and thereafter to be discovered, might be provided with names at once convenient and suggestive. This system, slightly modified and considerably extended—to accommodate the yet more widely expanding needs of the science—affords the basis of the chemical language of to-day.

A Few Principles of Chemical Language.

It is proposed to explain here a few of the first principles of chemical nomenclature and notation; that is, to present a few of the rules by which significant and useful *names and symbols* are provided. These will be found to meet the wants of hitherto inaccurately known substances, and even of those formerly unknown.

First; the names of elementary substances long known are retained. Thus, gold and silver are metals that were known in the earliest historical periods, if not in prehistoric times; their names, therefore, still remain in use.

Second: the discoverers of new elementary substances assign the names. In so doing, they usually invent a name that suggests some *fact* connected with the substance itself. Thus the name nitrogen is derived from two Greek words (νίτρον, *nitron*, mineral alkali, and γεννάω, *gennao*, I produce) carrying the suggestion that the gas is one of the constituents of nitre. The name hydrogen is is derived from two Greek words (ὕδωρ, *hydor*, water, and γεννάω, *gennao*, I produce,) indicating that wherever water exists, hydrogen is an essential constituent of it. So the name chlorine is derived from a Greek word (χλωρός, *chloros*, green,) which reminds the chemist of the fact that chlorine gas possesses a greenish color. The substance oxygen, however, was named in a different manner. Its name is derived from two Greek words (ὀξύς, *oxys*, acid, and γεννάω, *gennao*, I produce,) signifying a generator of acids. It appears then, that in this case the name is based, not on an easily verified fact, but upon a *theory*, current when oxygen was discovered, of the action of the substance in question. In a certain sense a name thus formed may be considered ill-advised. Thus in the case in hand it has turned out that while oxygen is a constituent of a majority of known acids, it is not essentially an acidifying substance: many acids are known that contain no oxygen at all, and again there are a multitude of compounds containing oxygen that are not acids in any proper sense.

Third; newly discovered metals are usually given names which, while they suggest some property of the substance, have in addition the termination, um. Thus the metal thallium derives its name from a Greek word (θάλλος, *thallos*, a green twig,) which carries the suggestion of the fact that thallium and its compounds when highly heated evolve light of a delicate green color. Again caesium, a newly discovered metal,

ANTOINE LAURENT LAVOISIER:
Born in Paris, August 26th, 1743; died on the scaffold in Paris, May 8th, 1794.

has a name **derived from a Latin word** (*caesius*, blue,) which refers to the fact **that** caesium and its compounds when highly heated afford light of a blue color. The termination *um* is used for **metals, after** the analogy of the Latin language which usually has its names **of** metals end in *um*. Indeed the chemist often makes use of the Latin names of even those metals that have been long known by more familiar ones. Thus for gold the Latin word *aurum* is used, for silver the **Latin** word *argentum*, for lead the Latin word *plumbum* ; it will be seen later that slightly modified forms of these names are very frequently employed when compounds **of these** metals **are** to be designated.

Symbols used for Atoms.

Each **elementary substance, or,** strictly speaking, **the minute** quantity **of it represented by the term one atom, may** be designated **in** brief by a special letter or short group **of letters** called the symbol. The usual symbol is the initial letter of the native or the Latin name of the substance. Upon examining the list of elementary substances at page 13, it will be seen that the following nine of the names begin with the letter *c ;* of course, therefore, **in eight** cases at least, the symbol must contain **an** additional distinguishing **letter :**

Accordingly C indicates one atom **of Carbon ;**
Ca " " **Calcium ;**
Cd " " **Cadmium ;**
Ce " " Cerium ;
Cl " " Chlorine ;
Co " " Cobalt ;
Cr " " Chromium ;
Cs " " Caesium ;
Cu " " Copper (Latin word *cuprum*).

It also appears that in the case of metals, like iron and copper, known to the ancients the symbols used are derived from the Latin names. The use of these symbols made from letters—and therefore called literal symbols, from the Latin word *litera*, a letter—will become apparent as the reader advances; but it is easily perceived that they afford a convenient abridgement of the longer titles of the elements.

The use of literal symbols as an abridgement of the chemical nomenclature was first proposed by Berzelius, a Swedish chemist, whose eminence in every branch of the science was such that the suggestion here referred to constitutes one of the least of the many and substantial grounds on which his fame rests.

READING REFERENCES.

Alchemy.

Rodwell, G. F.—The Birth of Chemistry. London. 1874.

Draper, J. C.—Amer. Chemist, v, 1.

Mackay, Charles.—Memoirs of Extraordinary Popular Delusions. 2 v. London. 1869. i, 93.

Berzelius.

Wöhler, F.—Early Recollections of Berzelius. Am. Chemist. vi, 131.

Chemistry, History of

Thomson, Thomas.—History of Chemistry, 2 v. London. 1830.

Hoefer, F.—Histoire de la Physique et de la Chimie. Paris. 1872.

Kopp, Hermann.—Geschichte der Chemie. 4Th. Braunschweig. 1843.

——Die Entwickelung der Chemie in der neueren Zeit. München. 1873.

Bolton, H. C.—Chem. News. xxxii, 36, 56, 68.

Liebig, J. v.—Familiar Letters on Chemistry.

JONS JAKOB BERZELIUS:
Born in East Gothland (in Sweden), Aug. 20, 1799; died Aug. 7, 1848.
(25)

Chemistry, History of

Whewell, Wm.—History of the Inductive Sciences. 2 v. New York. 1875. ii, 259.

Lavoisier,

Thomson, Thomas.—History of Chemistry. 2 v. London. 1830. ii, 75.

Figuier, L.—Vies des Savants Illustres du xviii siècle, 444.

Brougham, H.—Lives of Philosophers of the time of George III. Edinburgh, 1872. 290.

Nomenclature.

Morveau, Guyton de.—Memoire sur les denominations chimiques, la necessité d'en perfectionner le système, les règles pour y parvenir, suivi d'un tableau d'une nomenclature chimique; Dijon. 1782.

Lavoisier, de Morveau, Fourcroy, Baumé, Hassenfratz, Adet and others. Méthode de nomenclature chimique. Paris. 1787 (The Boston Athenæum Library contains a copy of this work.)

CHAPTER IV.

CLASSIFICATION OF THE ELEMENTARY SUBSTANCES.

N speaking of the elementary substances some of them have been referred to as metals. What then is the exact idea conveyed by this designating term? Everyone can readily picture in his mind some metal or metals like gold, silver, tin, zinc and others that have certain common characteristics, such as great weight, and the peculiar brilliancy and power of reflecting light which is described as metallic lustre. Another well marked and widely recognized characteristic at once thought of is the facility with which the substances ordinarily known as metals may be beaten or rolled into thin layers. This property, called malleability, (a word derived from the Latin word *malleus*, a hammer,) is not possessed in any striking degree by substances other than metals. Thus sulphur is not malleable : quite the contrary, it is very brittle. Charcoal, which consists mostly of the elementary substance called carbon, is likewise not malleable ; neither of these last two substances would be likely to be considered by even an ordinary observer as metals. In fact they are classed as non-metals by the chemist. This division of the elementary substances into metals and non-metals is dwelt upon, not because it can be called a very important one, but because it is widely used in works on chemistry

and because in deciding to which of these two classes a given substance belongs, ultimate dependence must be placed upon its chemical characteristics rather than upon its mere mechanical properties.

The Meanings Associated with the Term Metal.

The principal properties referred to are best presented in three groups.

First. Metallic Properties Associated with Mechanical Relations.

An elementary substance accepted as a metal must possess the property of existing in a solid condition; a weight rather greater than that of most well-known substances; considerable hardness, malleability, ductility (that is the capability of being drawn out into fine wire).

Second. Metallic Properties Associated with Physical Relations.

A metal should possess the metallic lustre; the power called opacity, by reason of which it does not allow light to pass through it; the noticeable capability of allowing heat to flow in it, called the power of conducting heat; the ready capacity for allowing the electric current to flow in it, called good conducting power for electricity.

Third. Metallic Properties Associated with Chemical Relations.

A metal should possess the power and the tendency to readily form a chemical union with oxygen; the chemical power to act upon compounds containing hydrogen, in such a way as to turn the hydrogen out, and take its place in the old compound and thus form a new one; the relationship towards the electric current such that when the element is subjected to the galvanic battery, it tends to gather about the negative pole—in consequence of which characteristic it is usually called electro-positive.

But while no known metal appears to possess the entire range of properties with which in thought the ideal one is endowed, every substance classified as a metal should possess many of them.

An illustration of what has been said may be found in metallic mercury. From the fact that under ordinary conditions it is a liquid it is plain that mercury must lack certain of the metallic properties referred to; that is, it does not possess the solid form, it does not possess hardness, it does not possess malleability, it does not possess ductility. Yet if it is cooled to a low temperature—about forty degrees below zero—it freezes, in other words becomes solid; then it possesses many of the distinctly metallic features that it necessarily lacks when in the ordinary liquid condition. Of course this liquid condition is a mere incidental circumstance, due to the temperature which ordinarily prevails upon our earth. If our ordinary temperature were slightly lower than forty degrees below zero, mercury would then be commonly known as a solid, hard, lustrous, heavy, malleable metal—capable of course of melting with a slight accession of heat.

As a further illustration, in a somewhat different direction, mention may be made of the metal lithium. This substance cannot be called heavy, since it is lighter than water; indeed it is the lightest solid known. But on the other hand it possesses in a striking degree those chemical features of metals, such as strong affinity for oxygen and tendency to combine with it, which have already been detailed in our definition of the ideal metal.

The Term Non-metal.

The term non-metal is suggestive of a negative idea, and not of any definite or positive one. In fact it is intended to intimate that elementary substances of this class are those which do not properly belong to the other. Sulphur and carbon have been already alluded to as examples of non-metals; other non-metals, such as oxygen, hydrogen, nitrogen and chlorine among the gases, bromine, a liquid, and iodine, antimony, phosphorus among solids, are far less familiarly known to most persons.

READING REFERENCES.

Elements, Classification of
   Williamson, A. W.—Jour. of Chem. Soc. of London.   xvii, 211.

Chemical Theory.
   Cooke, Josiah P., Jr.—The New Chemistry. New York. 1874.
   Remsen, Ira.—Principles of Theoretical Chemistry. Philadelphia. 1883.
   Tilden, William A.—Introduction to the Study of Chemical Philosophy.
   Wurtz, Ad. (Henry Watts, Tr.)—History of Chemical Theory. London. 1869.

CHAPTER V.

COMPOUND SUBSTANCES.

N a previous chapter a list of elementary substances has been given. All other matters known are compounds. From what has been said already, it is evident that the compounds are very numerous, indeed that there is practically no limit to the number of possible ones. These compounds are all made up by the union of elementary substances in obedience to the peculiar chemical forces that reside within them. Some compounds have only two kinds of elements: they are called binaries. Some compounds have three kinds of elements: they are called ternaries. Other compounds may have four, five, six, or even more kinds of elements grouped together to form one sort of substance. In this place reference will be made principally to binaries and ternaries—that is to the compounds of the simpler forms of constitution.

Examples of Binary Compounds.

In discussing binaries it will be well to give at the outset three or four examples for the purpose of illustration.

First. The gas known as hydrogen and the gas known as chlorine have the power of combining chemically and producing an entirely new compound, a compound different from hydrogen and different from chlorine yet containing portions of each of them. This compound is a binary since it consists of but two kinds of elements. It has several names, one of which is *hydric chloride*. The chemist frequently represents what is evidently the smallest possible quantity of this substance, and also its exact composition, by the expression

$$H\,Cl.$$

It is plain that this expression means a minute portion of substance formed by the union of one atom of hydrogen, (expressed by H,) and one atom of chlorine, (expressed by Cl).

Second. When sulphur burns in the air, it produces a blue flame. At the same time a new and peculiar gas is formed which is easily recognized by its choking odor, similar to that given off by a burning sulphur match. Now this odor is one of the properties of a new compound that has been formed: a compound different from sulphur, different from oxygen, yet containing them both and produced by the union of them. The compound is a binary because it contains but two kinds of elements. It is called *sulphur dioxide*. The name is intended to suggest that there are two atoms of oxygen to one of sulphur in the compound. This idea is further conveyed by the abridged system of notation so commonly used by chemists. Thus by this system the smallest possible quantity of the compound in question is expressed as follows,

$$SO_2.$$

In this expression it is very plain that S stands for one atom of sulphur, and O_2 for two atoms of oxygen.

Third. But sulphur may be made to combine with a still larger amount of oxygen than it takes when it simply burns in the air. Then it forms a compound called *sulphur trioxide*. This is still a binary, since it contains nothing but sulphur and oxygen, that is only two elementary substances.

Expressed in the briefer form the smallest quantity of this compound may be represented by the formula,

$$SO_3$$

This expression means a compound arising from the union of one atom of sulphur and three atoms of oxygen.

Fourth. When lead is heated to the melting point it is observed to become coated with a constantly increasing mass of a kind of ashes. A pound of the lead when heated in this way produces considerably more than a pound of dross. The formation of this dross is explained by the fact that when lead is heated it really burns, though of course the rapidity of the burning depends upon the amounts of heat and air to which the lead is subjected. Evidently the lead, in burning, has something added to itself. That something is a gas which is ever present in the atmosphere and which is called oxygen. The dross is a chemical compound of lead and oxygen. It is called *plumbic oxide*, and its smallest quantity is represented by the formula,

$$PbO.$$

In this formula it is easy to see that Pb stands for an atom of lead (whose Latin name is *plumbum*), and O for an atom of oxygen. The dross then is a binary compound.

A multitude of such examples of binary compounds might be given; probably those already cited are sufficient for the present. It will be advantageous to the reader to carefully learn the names and the formulas of the binary compounds thus far given, since they are selected examples which may be used again further on.

Examples of Ternary Compounds.

The ternary compounds are those which consist of three kinds of elements; of course they are more complicated in structure than the binaries.

COMPOUND SUBSTANCES.

This fact, however, must not deter the reader from the attempt to understand them at the outset, for the principal ternaries are acids and salts, and everyone knows that acids and salts are among the most important compounds which the chemist has to employ.

As examples of ternary acids mention will be made of two of the principal ones used by the chemist.

And first, *nitric acid* is a compound of hydrogen, nitrogen and oxygen. The formula of the smallest individual portion of it is

$$HNO_3$$

These letters signify that nitric acid contains one atom of hydrogen, combined with one atom of nitrogen and three atoms of oxygen. Now this nitric acid forms a great many salts. A simple example may be found in that one containing silver. Thus when nitric acid and silver are warmed together, either a part or the whole of the silver dissolves. A new substance is produced which is commonly called nitrate of silver. By the chemist it is oftener called argentic nitrate. Its solution may be dried into the form of a white crystalline substance, one that has long been accepted as a member of the class of salts. Its formula is

$$AgNO_3.$$

It is plain that this last formula is employed as a short way of expressing that the salt is a compound of more than one kind of element—of three kinds in fact—and that these elements are in the proportions of one atom of silver, one atom of nitrogen, and three atoms of oxygen.

Promise was made to refer to two important acids; *sulphuric acid* is the second one. Commercially this substance is by far the most important of all the acids. Indeed its manufacture is one branch of the greatest chemical industry devised by man—the alkali trade. Evidently it is important that the chemist should be thoroughly acquainted with sulphuric acid, with its composition, its formula, its way of chemically acting on other substances, and the things or products that it gives rise to when it has opportunity so to act. Now sulphuric acid has the formula

$$H_2SO_4$$

This formula means that sulphuric acid is a ternary, being made up of three different kinds of elements, namely two atoms of hydrogen, one atom of sulphur and four atoms of oxygen.

Further sulphuric acid forms a multitude of salts. Thus it forms one containing silver. This is commonly called sulphate of silver, though the chemist generally calls it argentic sulphate. The formula of argentic sulphate is

$$Ag_2SO_4$$

When this formula is firmly acquired by the reader so that he can readily compare it with others already mentioned, a certain simple and distinct relationship may be traced. Thus comparing

Argentic sulphate, Ag_2SO_4
with Sulphuric acid, H_2SO_4

it is evident that in the one, two atoms of silver have taken the places of two atoms of hydrogen that appeared in the other. And such is usually the case: when silver takes the place of hydrogen, it does so, atom for atom. Indeed argentic nitrate, $AgNO_3$, already described, illustrates this fact. It is a compound product closely related to nitric acid, HNO_3, the only difference of construction being that here also one atom of silver has taken the place of one atom of hydrogen.

The Purpose of this Chapter.

The purpose of this chapter has been to suggest a few facts respecting the nature of chemical compounds and also to show how the science of chemistry employs its peculiar language both in its longer and shorter forms. This language is very comprehensive. In fact it is too elaborate for full explanation here. The plan contemplated is to give at this point a few hints as to its nature and scope, and to develop it only so far as may be necessary to the succeeding stages of our progress.

It is proper to suggest at this point that no single scientific man —nor society of them—can *enforce* the use of any particular words upon the great body of chemists. For this reason, as well as for others, there still prevails the use of different chemical names for the same substance. Thus the compound of hydrogen and chlorine first referred to as represented by the formula H Cl, has at least four widely used names: *first*, a name merely suggestive of its component parts, that is hydric chloride; *second*, names which suggests something in addition to its component parts, namely that it is an acid, thus it is called both chlorohydric acid and hydrochloric acid; *third*, an old fashioned name, which still retains its hold upon the commercial world, namely, muriatic acid.

This same general principle applies to a great many other substances, and while it is true that it thus increases the number of names in the chemical language, it likewise incidentally enriches that language. For in many cases it has come to pass, little by little, that these different names are appropriated to slightly differing forms of the same substance, and so the name employed often conveys to the intelligent chemist as definite a shade of meaning as do the different synonyms used in the descriptions of every day affairs by any accomplished author. A single example will elucidate this point. The term oil of vitriol would usually be defined as meaning sulphuric acid. But the words sulphuric acid convey, strictly speaking, the same meaning as the formula

$$H_2SO_4$$

This latter substance, however, is of very rare occurrence alone: it is usually associated with varying quantities of water, and is then spoken of as sulphuric acid of varying degrees of dilution. Now in commerce the term oil of vitriol has come to be appropriated exclusively to that dilution consisting of about

                89 per cent. of sulphuric acid, H_2SO_4
                with 11 per cent. of water, H_2O,
both taken by weight.

READING REFERENCES.

Notation, Chemical

    Frankland, E.—Experimental Researches in Chemistry. 3. London. 1877.

    Williamson, A. W.—Jour. of Chem. Soc. of London. xvii, 421.

    Frankland, E.—*Loc. cit.* xix, 372.

    Madan, H. G.—*Loc. cit.* xxiii, 22.

    Council Chem. Soc. of London.—Instructions to Abstractors from Current Publications. (1879.) Chem. News. xlvii, 15.

CHAPTER VI.

THE CONSTRUCTION OF SUBSTANCES.

N order to understand the chemical construction of substances it is necessary to consider three terms much used by the chemist: these terms are,
Mass,
Molecule,
Atom,
Evidently the words relate to three grades of magnitude in which matter is capable of existing; it is equally plain that of the series the mass represents the largest individual portion of substance, and the atom the smallest, while the molecule represents the intermediate one.

The Chemical Use of the Term Mass.

Whoever looks about him sees substances existing in masses. This is true of vast mountain chains and equally true of the smallest grains of matter that are recognized as the humblest components of those peaks.

But the smallest of these *visible* masses is made up of particles still more minute—yet perhaps of precisely the same kind. For the chemist possesses means of subdivision of substances by which he may make them into minute fractional parts that are measurable and are all just alike, and he may continue this process long after the portions have sunk below the reach and range of ordinary vision. A lump of pure sugar as big as a cubic inch may be *mechanically* divided by any one into many smaller ones, each little one being easily recognized by the ordinary senses as possessing the sweetness, the crystalline construction, the whiteness, the solidity, the brilliancy, the power of dissolving in water and indeed a great many other well-known characteristics that pertain to sugar. But the chemist is able to continue the subdivision of the sugar much further. This he does by recourse to processes not exactly mechanical though closely allied to them: by processes often called *physical* as distinguished from purely mechanical ones. He may thus reduce the sugar to fragments of such extreme minuteness that while they do not impress our senses as larger portions do, yet each fragment is capable of displaying to a competent scientific observer the certain and sure chemical properties that always belong to sugar, whether in large lumps or in small ones, and which in fact belong to nothing but sugar.

Speaking generally, all particles producible by mechanical subdivision are masses, while the same is true of most particles producible by physical subdivision.

The Chemical Use of the Term Molecule.

But there is a point where any attempt at further scientific subdivision results in a new and startling change: at this stage the last individual that can properly be called sugar is dissected and loses entirely the characteristics of sugar. The fragments produced by the wreck of the last particle are of a new kind. They are

portions of carbon,
portions of hydrogen,
portions of oxygen.

This last particle of sugar **is separated into its** ultimate constituents only by *chemical* processes. **This last particle before it is** broken up is called the *molecule*, the word meaning a little **portion.** The single individual thing it refers to **cannot be detected by the eye, nor can** it be **in any way appreciated except by** scientific means. **But a** chemical change of *the last multitude of molecules at once*, is practicable to everybody, and it is to a certain extent recognized by every one who heats sugar until **it turns to a charred mass.** This charred **mass is** mainly carbon—one **of the components of the** now ruined sugar—and it **is very unlike sugar in every way. The chemist** can show that when sugar **is charred,** the oxygen **and the** hydrogen go **off** mostly in the form of gases **or vapors, and that on** this account they escape detection **at the** hands of **all ordinary observers.**

The Chemical Use of the Term Atom.

It appears **then that the chemist is able to subdivide** molecules into smaller parts. **But he finds that further division is at a** certain stage forbidden him. **He can take the** oxygen out **of the sugar, but he** cannot **take** anything but oxygen **out** of oxygen**; he can take** hydrogen out of **sugar, but he** cannot take anything **but hydrogen out of hydrogen; he can take** carbon out of sugar, **but he cannot take anything but carbon out of carbon.**

As a result of all chemical study **of common sugar the chemist has** fixed upon the following as expressing most **closely the facts as he knows them :**

The formula
of one molecule of pure cane **sugar, is**
$C_{12}H_{22}O_{11}$

The chemical formula of any substance expresses much more than the **reader would at first imagine.** Thus the formula $C_{12}H_{22}O_{11}$ conveys at once to the chemist a series of facts, some of which may be amplified as follows :—one molecule of sugar contains three kinds of substance : carbon, hydrogen and oxygen :—each of these kinds of matter exists in the molecule in separate minute portions such as in the present state of human knowledge are divisible only in a limited way : thus the carbon of one molecule of sugar is divisible into twelve parts, *and no further*;

the hydrogen of one molecule of sugar is divisible into twenty-two parts, *and no further*;

the oxygen of one molecule of sugar is divisible into eleven parts, *and no further*.

Definition of the Term Atom.

Now at last the atom has been reached. It is that portion of any kind of matter that is to human beings indivisible *in fact*. It has already been stated that there are only sixty-six different kinds of atoms, it appears then that there are only sixty-six kinds of matter that at present cannot be chemically subdivided into different components.

It is true that some persons consider that certain intricate chemical processes suggest that what have been here called indivisible atoms are themselves really capable of yet further decomposition. Without attempting here to sustain or to demolish this proposition or to say what the future of chemical investigation may reveal, it may be safely remarked that adequate proof has not yet been offered of the ability of any one to successfully accomplish a decomposition of the sixty-six atoms enumerated.

Plainly then just as bricks may be made into a building, and a series of buildings may make a city, and a series of cities may exist in a state, so atoms may combine together to form molecules, and molecules may cohere together to form a mass, and visible masses may be placed side by side and give rise to the ordinary objects recognized about us. True the comparison suggested is not strictly carried out in all particulars. But the difficulty not a serious one ; for a city *might* contain a multitude of houses each one so similar that no difference could be distinguished between them, just as a mass of sugar does in fact contain molecules of which each one is so like its neighbor that the most refined chemical methods discover no difference between them. Again these same houses *might* be composed of combinations of brick and other materials differing among themselves, but closely corresponding in every house. So the molecule of sugar does contain atoms of carbon, hydrogen and oxygen, the atoms of one kind differing distinctly and absolutely from the atoms of the other kind.

But here the parallelism seems to cease. For while all bricks and other components of a building are capable of being split into smaller portions, the atoms composing the molecule are found by the chemist to be absolutely indivisible in the present state of knowledge.

Employing still further the illustration already in hand it may be added that just as the walls of a dwelling might contain bricks either of the same kind as to their color, shape and weight, or else differing in these or other respects, so a molecule may be a little group of atoms of the same kind, or it may be a group of atoms of different kinds. Thus the hydrogen gas molecule is composed of two atoms each just alike, and each being hydrogen. This molecule is represented by the formula

$$H_2 \text{ or } H-H.$$

So a molecule of chlorine gas is composed of two atoms each just alike and each being chlorine. This molecule is represented by the formula

$$Cl_2 \text{ or } Cl-Cl.$$

Everything Built up of Atoms.

Now each of the sixty-six elementary substances has molecules composed of atoms, and each molecule of a given element is composed of atoms of the same kind. And further all the vast and countless myriad of compound substances, whether buried in the heart of the solid earth, whether drifting in the wandering courses of the ocean's currents, whether floating in the airy mass which is wrapped about our globe, whether components of distant planets of unknown constituents, whether parts of the seething mass which pours its volcanic torrents millions of miles out from the surface of our central sun—all these substances are constructed, so far as we know, of inconceivably minute atoms of varying kinds bound together by chemical attraction into molecules, the molecules being piled one upon another into those masses whose reaction our dull senses can appreciate.

Atoms and Molecules Manifest Chemical Affinity.

But to the chemist, atoms, molecules and masses possess an interest of another kind. Each atom and each molecule is endowed with an invisible, occult power called *chemical affinity*. This power acts like an unseen spirit possessed of likes and dislikes. By reason of it an atom of hydrogen for example instantly binds itself to an atom of chlorine whenever opportunity offers, but will never, even under the most favorable circumstances, combine with an atom of gold.

Finally this attractive force is a kind of energy of which no true explanation can be offered. All that human beings can do is to attentively study it as it manifests itself in the relations of elementary substances and compound substances one toward another. Indeed one of the principal offices of chemistry is to study these relationships as they develop. It is the multitude of possible relationships and actions of which the numberless substances known are capable, that gives to chemistry its great scope and variety and that makes it such a vast field for experiment, for discovery of facts, and for industrial application of them.

READING REFERENCES.

Atoms and Molecules.
    Barker, G. F.—Amer. Chemist. Nov. 1876. p. ms. 164.
    Stoney, Johnstone.—Phil. Mag. xxxvi, 141.
    Clerk-Maxwell J.—Encyclopædia Britannica, *article* Atoms.
    ——Theory of Heat. New York. 1872.
    Thopson, Sir Wm.—*Nature.* Mch. 1870.
    Tait, P. G.—Recent Advances in Physical Science. London. 1876. p. 283.
    Mayer, A. M.—Lecture Notes on Physics. p. 52.
    Cooke, J. P.—The New Chemistry. New York. 1881. pp. 29–43.
    ——American Cyclopædia, *article Molecule.*

CHAPTER VII.

HOW CHEMICAL AFFINITY WORKS.

IT is an interesting fact that elements and compounds manifest an exceedingly great variety of tendencies to combination. Fragments of matter, so small that no eye perceives them, have, wrapped up in themselves, a multitude of determinate powers. A given atom as of lead, for example, will very readily combine with oxygen and with some other substances, but it seems to absolutely refuse to combine with nitrogen. So hydrogen will combine very readily with chlorine and with many other elements, but it refuses to form any union with silver and with many other elementary substances. It cannot be called a whim that determines the kind of element or its amount that a certain substance will combine with, though the likes and dislikes of atoms are in this respect exceedingly marked and even incomprehensible. But however impossible it may be *to explain* an element's friendly or unfriendly deportment toward another, it is possible in each case to learn the facts with certainty, for each atom possesses its true individuality and is always constant and consistent in its affinities and hates.

It is the purpose of this chapter to present in an orderly manner some of the peculiarities of this mysterious power of chemical affinity.

First. Each Kind of Atom Has Its Peculiar Chemical Affinities.

Chemical affinity seems to reside within the atom as a permanent, ever present and guiding energy. Thus while iron oxidizes readily—that is manifests under a multitude of common conditions a willingness to combine with oxygen and form a new compound called oxide of iron, and well-known under the name of iron-rust—gold on the other hand oxidizes unwillingly; indeed in order to get it to combine with oxygen it must be coaxed by means of circuitous and carefully planned devices. But these atoms are always consistent in their action, for iron under any and every condition oxidizes more readily than gold does.

Second. Chemical Affinity Acts Only Under Favorable Conditions.

While chemical action often works with most intense energy it does so only when certain outside and incidental conditions are favorable. Thus carbon has under certain conditions an affinity for oxygen and manifests its tendency to combination, with an intensity that is scarcely surpassed. In order however to awaken and vivify the dormant inclination it must be stimulated by certain definite and favorable conditions; the most important of these conditions is a certain amount of warmth. The stores of fuel in our cellars—the coal and the wood and all other combustible things—are surrounded by great quantities of oxygen which winds its way, with every slightest stir of the mobile air, in and out through all the crevices that the fuel affords, passing continually in the immediate neighborhood of ample quantities of atoms of carbon. But it does not ordinarily unite with them. Subject the whole or any portion of these combustible things to a slight rise in temperature—then the atoms of oxygen and the atoms of carbon seem to arouse themselves from repose: they unite in friendly and firm grasp, and what is called chemical union takes place. To the ordinary observer the heat that is produced is the most notable sign of this kind of combination. The chemist, however, discovers a yet more conclusive evidence, for he finds that several kinds of new molecules have been produced; one of these kinds, for example, is expressible by the name carbon dioxide and by the formula CO_2. Evidently this formula means that each atom of carbon has united with two atoms of oxygen. In this familiar example heat is the agency that stimulates the atoms to a display of the chemical force that previously was slumbering within them.

Light and the electric current and the vital forces of animals and plants, though acting in a manner less familiar to us, are energizers of chemical affinity and all have their proper influence to make atoms join in union; indeed in some cases they make atoms burst from each others bonds and fly away to more congenial conditions.

Third. Each Atom Has a Certain Equivalence or Atom-Fixing Power.

The chemist also recognizes each atom as possessing certain peculiar numerical preferences in its combinations; a manifestation of chemical affinity called equivalence. Thus when carbon burns in a stove, by reason of the air passing by it on its way to the chimney, it seizes upon some of the oxygen atoms and *binds a definite number of them* to itself. If there is much air, each atom of carbon of the millions present, picks out two atoms of oxygen from the air; if there is but little air, each atom of carbon has to be satisfied with one atom of oxygen. Now in these two cases of course different substances are formed. The first, whose composition is represented by the formula CO_2, has already been spoken of as carbon dioxide. To the other, whose composition is represented by the formula CO, is applied the name carbon monoxide. Here then we see that *the same atom may sometimes combine with two atoms of oxygen and sometimes with only one.*

Further the chemist knows four simple and familiar compounds whose molecules illustrate very strikingly the difference of equivalence of *different atoms*. These compounds are the following:

Chlorohydric acid	(Hydric chloride),	HCl	or H–Cl
Water	(Hydric oxide),	H_2O	or $\left.\begin{array}{l}H-\\H-\end{array}\right\}O$
Ammonia gas	(Hydric nitride),	H_3N	or $\left.\begin{array}{l}H-\\H-\\H-\end{array}\right\}N$
Marsh gas	(Hydric carbide),	H_4C	or $\left.\begin{array}{l}H-\\H-\\H-\\H-\end{array}\right\}C$

It has beeen found advisable to adopt the atom of hydrogen as the standard of equivalence or atom-fixing power. It is plain that by this method of comparison the atom chlorine may be said to have the equivalence *one*, since it combines with one atom of hydrogen. And so the atom oxygen may be said to have the equivalence *two*, since it combines with two atoms of hydrogen. And the atom nitrogen may be said to have the equivalence *three*, since it combines with three atoms of hydrogen. And the atom carbon may be said to have the equivalence *four*, since it combines with four atoms of hydrogen.

The language of chemistry sometimes presents the same observed facts in a slightly different form, somewhat as follows: Chlorine is said to have one point of attraction and is called a monad (a term derived from the Greek word μονάς, *monas*, a unit). Oxygen is said to have two points of attraction and is called a dyad (a term derived from the Greek root δυάς, *dyas*, two). Nitrogen is said to have three points of attraction and is called a triad (a term derived from the Greek word τριάς, *trias*, a group of three). Carbon is said to have four points of attraction and is called a tetrad (a term derived from the Greek word τετράς, *tetras*, four).

While hydrogen as the basis of the system has the uniform equivalence one, and is always a monad, and oxygen its close friend and ally has always the equivalence two, and is always a dyad, most other elements have some variety of equivalence. Thus chlorine has at different times different equivalences, sometimes *one*, sometimes *three*, or *five*, or *seven*. So nitrogen has at different times the different equivalences, *one*, *three*, *five*. So carbon has sometimes an equivalence *two*, sometimes *four*.

Fourth. Chemical Changes Neither Create nor Destroy Matter.

When chemical changes are produced by reason of the action of chemical affinity, there is never either gain in weight or loss in weight. In other words there is no creation of matter and no destruction of it. In former times, people who observed the disappearance of solid matter when charcoal burns, thought that the substance was destroyed—partly if not wholly. The modern chemist finds, however, that the carbon is only turned into the form of an invisible gas, and that by the use of appropriate appliances he can find the weight of this gas, and compare it with that of the carbon producing it. In the combustion of carbon the chemical change is represented by the following equation :

C	+	O_2	=	CO_2
One atom of		Two atoms of		One molecule of
Carbon		Oxygen		Carbon di-oxide.
12		32		44
parts by weight.		parts by weight.		parts by weight
44				44

This equation means that the chemist has discovered, by careful experiments, that when any twelve parts by weight of carbon—say twelve pounds—are completely burned, they always unite with thirty-two corresponding parts of oxygen (in this case thirty-two pounds), and they produce forty-four parts by weight of carbon di-oxide (in this case forty-four pounds).

And so in all chemical changes the substances taking part—whether solid, liquid, or gaseous—may be weighed, and the sum of the weights of all the matters finally produced is just equal to the sum of the weights of the original factors.

Fifth. Chemical Changes are Often Attended by Displays of Force.

In many chemical changes the union of the atoms is attended with the production of heat, or electricity, or some other form of energy. Now it is a law derived from modern discoveries that the amount of energy given out by any chemical union is fixed and invariable, and that it is just the same in amount as the quantity of that kind of energy that is absorbed when such chemical action is reversed.

Sixth. Chemical Changes Produce Striking Results.

Each of the atoms of matter is in itself fixed and unchangeable and it possesses through all its varied combinations an inherent character which belongs to it and which no human being can permanently alter. But when atoms unite to build up either simple or complex molecules, the various original atomic characters are so blended and balanced and reinforced as to afford in the molecular product an entirely new and unexpected set of properties. An example of these principles is found in the union of copper, sulphur, oxygen and hydrogen. These substances may combine to form a new molecule which is called cupric sulphate, and which has the composition expressed by the formula

$$CuSO_4 + 5H_2O$$

Of the constituents of this molecule, copper is red, sulphur is yellow, oxygen is colorless, hydrogen is colorless; but when they unite the cupric sulphate formed is blue, that is its color is not that of either of its constituents, nor is it intermediate between them. There is simply a new and unexpected result, and one which in the present state of knowledge cannot be explained ; it can merely be recorded. And this example is only one of a myriad. Throughout nature chemical changes most marked—and to the human thought unexpected—arise from the union of familiar elementary substances.

Seventh. Chemical Atoms Unite in Obedience to Definite Law.

Careful chemical study of the way in which atoms combine has developed the following as a fundamental law of nature. The same chemical compound always contains the same kind and number of elementary atoms, and these atoms are united in the same proportions by weight. This law is a formal statement of facts similar to those already referred to in paragraphs *third* and *fourth* of this chapter. It does not therefore seem to call for further explanation at this point.

The Modern Atomic Theory.

The same chemical study which has developed the truth of the law just stated has also given rise to the modern atomic theory. The chemist is constrained to believe that matter is composed of ultimate indivisible particles called atoms. While these atoms are invisible to mortal eye even with the help of the finest known optical appliances, yet when their existence is once admitted this admission affords an explanation that is a satisfactory one, and indeed the only one that harmonizes with the multitude of observed chemical and physical laws.

This atomic theory, in its essential particulars, was suggested in the early part of this century by Dr. John Dalton, who was a teacher of mathematics in Manchester, England, and who died as recently as in 1844. Dalton found recreation in chemical experiments, and the mathematical turn of his mind led him to express the results of his chemical analyses in a new numerical form. Thus previous to his time it had been customary to

JOHN DALTON:
Born at Eaglesfield, (England,) Sept. 5th, 1766; died July 27th, 1844.

express the composition of all substances in the ordinary percentage form. Now Dalton found that *if some special weight was adopted as the unit*, a variety of new and previously concealed facts was revealed. The idea to be here conveyed is partially but perhaps sufficiently expressed by the following examples derived from the two compounds of carbon already referred to :

Composition of the Two Compounds of Carbon and Oxygen.

EXPRESSED IN PER CENTS.	
Carbon Monoxide (CO).	*Carbon Dioxide* (CO_2).
Carbon, 43 — per cent.	27 + per cent.
Oxygen, 57 + per cent.	73 — per cent.
100	100

EXPRESSED IN DALTON'S FORM.	
Carbon Monoxide (CO).	*Carbon Dioxide* (CO_2).
Carbon, 12 parts by weight,	12 parts by weight.
Oxygen, 16 parts by weight.	32 parts by weight.
28	44

In Dalton's expression it is at once evident that, as compared with the weight of carbon, the amount of oxygen in carbon dioxide is exactly twice what it is in carbon monoxide : but to the ordinary unmathematical mind this fact is buried in the percentage statement. Dalton's experiments with still other compounds gave him results showing a simplicity of relationships similar to that obtained from the carbon compounds just referred to. To his mind these facts suggested immediately the idea that an elementary substance is made up of atoms each of a determinate weight, and that these atoms combine by wholes and not by fractional parts, and that although it is impossible to weigh any atom separately, yet *the weight ratios of a multitude of them that combine as wholes* express at once the weight ratios of the atoms themselves. He thus got the idea of atomic weights and constructed the first table of them. Since Dalton's first declaration of his atomic theory, the combining numbers of the different atoms have been studied by chemists with the most thoughtful care and the most painstaking methods known to modern science; and tables have been constructed showing the combining numbers which are believed also to be the true atomic weights for all the various elements thus far recognized.

READING REFERENCES.

Atomic Constitution of Bodies.
Saint-Venant.—Jour. of Chem. Soc. of London. xxx, pt. II, 472.

Atomic Philosophy.
——— Amer. Chemist. iii, 326.

Atomic Theory.
Williamson, A. W. (and others.)—Jour. of Chem. Soc. of London. xxii, 328, 433.

Wurtz, Ad.—The Atomic Theory. New York. 1881.

Atomic Volumes, Etc.
Avogadro.—Annales de Chimie et de Physique. 3 Ser. xiv, 330; xxix, 248.

Atoms, Vortex Theory of
Thomson, Sir Wm.—Phil. Mag. 1867.
Thompson, J. J.—Science. iii, 289.
Tait, P. G.—Recent Advances in Physical Science. London. 1876. p. 283.

Atomic Weights, Dalton's First Table of
Roscoe, H. E.—Chem. News. xxx, 266.

Chemical Operations, Calculus of
Brodie, B. C.—Jour. of Chem. Soc. of London. xxi, 367.

Dalton, John
Henry, W. C.—Life of Dalton. London. 1854.

Definite Proportions, Variabiliy in Law of
Boutlerow.—Silliman's Journal. 3d Ser. xxvi, 63.
Cooke, J. P.—loc. cit. 310.

Equivalents of the Elements.
Dumas, J.—Annales de Chimie et de Physique. 3 Sér. lv, 129.

Energy.
Stewart, Balfour.—The Conservation of Energy. New York. 1874.
Tait, P. G.—Recent Advances in Physical Science. London. 1876. Encyclopædia Britannica. vol. viii.

Gaseous and Liquid States of Matter.
Andrews, T.—Jour. of Chem. Soc. of London. xxiii, 74; xxx, pt II, 159.
Ramsey, W.—Jour. of Chem. Soc. of London. xlii, 156.

Matter, Constitution of
Ditte, A.—Annales de Chimie et de Physique. 5 Sér. x, 145.

Nomenclature of Salts.
Madan, H. G.—Jour. of Chem. Soc. of London. xxiii, 22.

CHAPTER VIII.

HYDROGEN.

HIS substance is one of the most interesting with which the chemist has to deal. On account of its chemical and physical properties, by reason of the many important compound substances into which it enters, by reason of the part it has played in the history of chemical progress, it is entitled to a large share of the student's attention.

Meaning of the Word Hydrogen.

The name hydrogen was applied to it some time later than the first recognition of the substance. The word is derived from two Greek words (ὕδωρ, *hydor*, water, and γεννάω, *gennao*, I form or produce), the word as a whole meaning *water former*. In fact hydrogen is in all water wherever that substance exists. That this is a very comprehensive expression appears when it is remembered that the atmosphere always contains water diffused through it in the form of invisible vapor even before that vapor is precipitated as the gentle dew, or the crystalline snow, or the streaming rain.

Again, water in seas and oceans, lakes and rivers, is the mantle of nearly three-fourths of the earth's surface. Every living being on the dry land, whether animal or vegetable, contains large quantities of water in its structure: the blood of the higher animals is nearly nine-tenths water.

While water is the principal substance containing hydrogen, this gas exists also as a constituent part of a great many other solid and liquid matters found in the earth.

Why Free Hydrogen is not Found in the Earth.

Hydrogen scarcely ever exists on our globe alone, that is in the free or uncombined condition. Indeed there are certain definite reasons why it should not. These are based mainly upon the very strong chemical affinity that hydrogen has for oxygen. Now, as has been declared already, the latter substance is the most abundant element in nature, and it exists in very large quantities in our atmosphere. Spread all over the surface of the earth then, the free oxygen of the air stands prepared to combine with hydrogen wherever the latter may be liberated. Such combination might not occur, it is true, unless initiated by influence of heat or some flame of fire; but owing to the constant agitation of the air by reason of uniform currents like trade winds, as well as those produced when the atmosphere is agitated by violent storms, any mixture of hydrogen and oxygen would be likely soon to come into contact with some flame or fire, and so these components would enter into combination. Thus hydrogen would not be likely to remain long uncombined even were it produced in considerable quantity by natural terrestrial operations.

The Discoverer of Hydrogen.

Hydrogen was first distinctly described and its properties as a special kind of gaseous matter clearly pointed out in the year 1766, by an English chemist, the Honorable Henry Cavendish. This philosopher, the son of Lord Charles Cavendish, and the grandson at once of the Duke of Devonshire and the Duke of Kent, is one of the most curious characters in the history of the natural sciences. He was of an exceptionally careful, thorough and pains-taking temper, which well fitted him for the scientific pursuits which were the prime objects of his thoughts. Sir Humphry Davy said of him: "The accuracy and beauty of his earlier labors have remained unimpaired amidst the progress of discovery, and their merits have been illustrated by discussion and exalted by time."

In addition to his possession of many special aptitudes for the exact studies to which he devoted his entire existence, it should be recognized that he lived at a period that was remarkably favorable to the pursuit of the natural sciences. The times, the state of knowledge, the condition of society all over Europe seemed to be ripe for this kind of progress, for in Scotland, in England, in France, in Germany, in Sweden there appeared experimenters of unsurpassed skill, and chemistry as a science had then its birth under most fortunate auspices.

Cavendish was very peculiar in his manners and habits, living in great seclusion and retirement and in the most simple and methodical manner; indeed his oddities attained for him the unenviable distinction of a place in a book devoted to the lives of English eccentrics. In that work, as well as in Dr. Wilson's life of him, are many amusing anecdotes of his way of life. One most remarkable episode was his inheritance of wealth. Though poor in his youth he was suddenly made rich in middle life by a bequest whose origin is scarcely known. M. Biot neatly described him as "le plus riche de tous les savants, et probablement aussi, le plus savant de tous les riches." He lived on however in as great seclusion as before, his chosen associates being his flasks and his thermometers. His millions made no observable impression upon his habits, notwithstanding at his death they made him the largest holder of the stock of the Bank of England.

Lord Brougham says that Cavendish probably uttered fewer words in the course of his life than any other man who ever lived to fourscore years,

JOSEPH BLACK, M. D.
Born in Bordeaux, in 1728; died in Edinburgh, Nov. 26, 1799.

not at all excepting the monks of La Trappe—who were bound to perpetual silence except in cases of absolute necessity.

Why Hydrogen was not Discovered Earlier.

Doubtless those prehistoric men who in earliest days looked about upon the face of the earth, curiously examing their heritage from the Creator, were familiar with water in its various forms. They must have prized its bland and refreshing powers and have learned many of its most important uses. But the idea that it is made up of more than one kind of substance or matter was not suspected until very recent times, and not proved until the masterly investigations of Cavendish clearly set forth the facts. Indeed the very idea of a chemical compound, that is of a substance as made up of inconceivably small portions of matter in a union of almost inconceivable intimacy, an idea very familiar to students of the present day, probably did not enter the minds even of those profound thinkers who suggested the earlier atomic philosophies. In fact the notion of chemical union is scarcely more than a century old.

Moreover, hydrogen is a gas, and *the notion of gas* is itself decidedly a modern one. It was first stated in well-defined form in the year 1752, by Dr. Joseph Black, professor in Glasgow and Edinburgh. Black clearly and conclusively demonstrated the existence of *airs* of a different kind from that familiar to us in our atmosphere. It is true Van Helmont and even others, fully one hundred years before Black's time, had known and stated more or less distinctly the existence of a gas or air different from that we breathe, but owing to a variety of circumstances these wonderful discoveries were allowed to relapse into forgetfulness. Thus the human race lost for a century much advantageous knowledge ; but probably the general social advancement of those times had not then prepared mankind for the benefits which the development of modern chemistry has conferred upon the present citizens of the world. Again, *experimenting* with gases was not well understood until about the year 1770, when Joseph Priestley invented that contrivance for manipulating them known as the pneumatic trough, for which no better substitute has yet been devised.

Further, in water—which has already been referred to as the most abundant and widely diffused compound of hydrogen—the partner elements are bound together by a chemical affinity that cannot be readily overcome. This intensity of attractive force between the constituent elements is therefore another reason why the true composition of water was so long an unsolved riddle and why hydrogen was not earlier recognized as a thing or kind of matter by itself, although in its principal compound—one of the most admirable gifts of the Creator to man—it was well-known from the first days of the human race.

How Hydrogen is Prepared.

Hydrogen may be obtained by the chemist in several ways:

First.—There is a method of directly tearing the elements composing water apart from each other. Considered theoretically this process is a most direct and simple one. In order to realize its results, however, advantage must be taken of the galvanic current. This force may be obtained readily it is true: thus in most cases where two metals, dipped in a liquid, are connected by a wire it is generated. But no one knows fully what the current is. The words galvanic current and voltaic current suggest the two investigators, Galvani and Volta, who were the pioneers in this field. but they give nothing that can be called an *explanation* of the wondrous, invisible, imponderable form of energy referred to. It is a force of an exceedingly interesting character and about which a certain considerable body of knowledge has been collected.

Among the variety of facts known about it is that one which relates to water; namely, when the poles or electrodes of a suitable galvanic battery are dipped into a vessel of water, bubbles of gas may be seen to flow freely from each of them. The gases may be collected in a vessel placed over the electrodes, but the experimenter may well beware of incautiously treating what has now been produced; he has obtained a mixture of oxygen and hydrogen from the original water, and these elements which he has rended apart from their more intimate union, are ready upon the approach of the smallest flame to rush into union again, with extraordinary violence, and in such a way as to produce a tremendous explosion. In the act of this explosion, therefore, water is again produced, first as expansive vapor, then condensible back to the liquid drops whence it came.

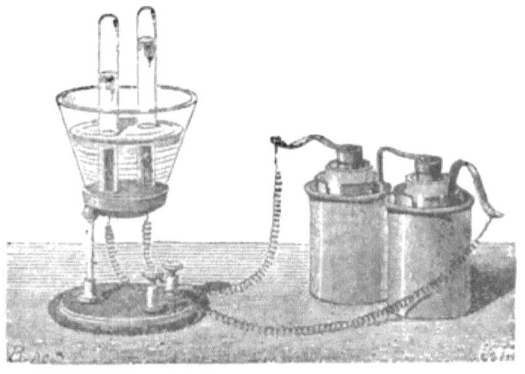

FIG. 6.—Apparatus for decomposition of water, (by action of two cells of the Bunsen galvanic battery,) and for collection of hydrogen and oxygen gases in separate receivers over the two electrodes of the battery.

If however the product from each electrode is collected *by itself in a separate tube*, the one gas is found to be very different from the other. The one is found to be hydrogen, the other oxygen. In accordance with the formula H_2O—which it has before been stated represents the composition of water—the hydrogen is found to be given off in a bulk or volume that is twice as great as that of the oxygen obtained at the same time from the same amount of water.

Second.—Hydrogen may be obtained by bringing into contact with water under proper conditions certain substances that have a very strong affinity for its oxygen and at the same time but little affinity for its hydrogen. Now every one is familiar with the fact that iron rusts readily in the air. The chemist can demonstrate that this rust is a compound of iron and oxygen. The union of these elements under ordinary conditions suggests at once that that union arises from an affinity between the iron and the oxygen. This affinity is much greater at high temperatures, for it is well known that iron rusts more violently when subjected to heat. These facts then are made use of for the purpose of withdrawing oxygen from water and thus forcing the hydrogen out in the free or uncombined condition so that it may be obtained and experimented upon.

To produce hydrogen by this method, there must be provided a long iron pipe which passes through a hot furnace; the pipe should contain frag-

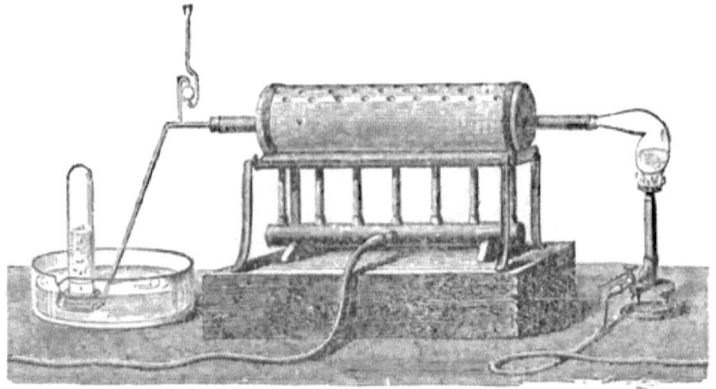

Fig. 7.—Apparatus for preparation of hydrogen gas. Steam, generated in the small retort, is conveyed through the tube placed in the gas-furnace; iron turnings within the tube being highly heated, decompose the water-vapor, which thereby evolves hydrogen. The liberated gas is collected in the little bell-glass.

ments of iron such as iron turnings, or iron filings, or pieces of iron wire. Then a current of steam must be passed through the pipe. The iron becomes red hot, and under these circumstances manifests more affinity for the oxygen of the steam than the hydrogen does. The iron then grasps the oxygen and holds it fast. As a result a peculiar kind of oxide of iron of a black color is produced. Its chemical formula is Fe_3O_4 and it is called by chemists ferroso-ferric oxide. The iron has now taken the place as a partner of the oxygen that the hydrogen formerly had. The hydrogen is thus cast out from its combination and is set free as an uncombined gas,

HYDROGEN.

in which liberated condition it is expelled at the end of the tube. The chemical action between the iron and the steam may be represented by the following equation :

Fe_3	+	$4H_2O$	=	Fe_3O_4	+	$4H_2$
Three atoms of Iron,		Four molecules of Water,		One molecule of Ferroso-ferric oxide,		Four molecules of Hydrogen,
168		72		232		8
parts by weight.		parts by weight.		parts by weight.		parts by weight.

$$240 \qquad\qquad\qquad 240$$

The gas produced as just described may be collected by adjusting a suitable tube in connection with the pipe containing the iron. When the gas is examined it is found to be in fact hydrogen. It will burn with a blue flame and perform all the various actions that acknowledged hydrogen will.

Third.—There are other metals, not known to the common every-day uses of life but still familiar to the chemist, which have far greater affinity for oxygen than iron has. One such metal **is that called sodium**. Its affinity for oxygen is so great that it **cannot be long preserved if exposed** to the air: a block or lump of it would, in a day **or two in the open air, turn entirely to a mass** of rust of sodium, that is oxide of sodium. This metal therefore is preserved by the chemist **in** bottles containing petroleum oil. **The oil keeps the air away from the metal ; moreover** the oil contains **no oxygen in** its composition as many **other liquids do. This metal** sodium though heavier than the oil is lighter than water. **If thrown upon water it floats.** But by virtue of its intense affinity for oxygen, it at the same time decomposes the water. It draws the oxygen to itself and it liberates the hydrogen. Some chemical skill is requisite in the performance of this apparently simple experiment, for occasionally the violent affinities involved set the sodium and the **hydrogen on** fire **and give rise to** dangerous explosions. When **properly conducted, however, the hydrogen** from this process may be **collected in a vessel and its various characteristics displayed.***

---

*Appleton's "Young Chemist," Philadelphia, Cowperthwait & Co. pp. 26, 27, 28.

Fourth.—The most common way of producing hydrogen is by bringing together sulphuric acid and zinc. The formula for sulphuric acid is H_2SO_4. Now the zinc has affinity for the compound radicle SO_4, known as the sulphuric acid radicle. The chemical change is represented by the following equation:

Zn	+	H_2SO_4	=	$ZnSO_4$	+	H_2
One atom of Zinc, 65 parts by weight		One molecule of Sulphuric acid, 98 parts by weight		One molecule of Zinc sulphate, 161 parts by weight		One molecule of Hydrogen, 2 parts by weight

$$163 \qquad\qquad 163$$

Here it is plain that by reason of its affinities the zinc has taken the place of the hydrogen—or the place which the hydrogen formerly held as related to the sulphuric acid radicle, SO_4,—and that the hydrogen thereby left without anything to combine with, appears as a free and uncombined substance. The hydrogen produced by this method can be readily collected and examined.

Perhaps it ought to be stated that neither of the processes thus far explained is likely to yield hydrogen in an absolutely pure condition. The various substances used are likely themselves to contain associated with them small amounts of other substances which give some impurity to the gas evolved.

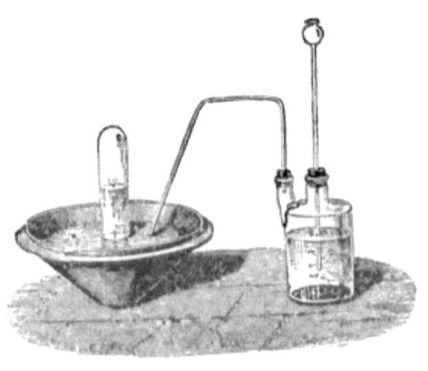

FIG. 8—Apparatus for production of hydrogen, by action of sulphuric acid on zinc, and for collection of the gas in a receiver.

The Powers and Properties Manifested by Hydrogen.

Hydrogen has been seen, from the explanation already given, to be a gas. Down to within a few years it resisted all attempts to liquify it. Chemists submitted it to intense cold and enormous pressure and to both these influences at the same time but without avail. Within a few years, however, by use of ampler resources and contrivances for the application of these condensing agencies, it has been brought down to the liquid and perhaps even to the solid state.

As a gas it is colorless, odorless, tasteless.

Bulk for bulk it is the lightest substance known in nature. Thus a quart of atmospheric air, light as it is, weighs over fourteen times as much as a quart of hydrogen. A cubic inch of gold weighs more than two hundred thousand times as much as a cubic inch of hydrogen. This lightness is properly illustrated by inflating a soap bubble with hydrogen rather than with air. When soap bubbles are filled with air they fall, unless indeed carried upward by a temporary current; but when filled with hydrogen they invariably rise with great rapidity. By reason of this great lightness hydrogen was formerly used for the inflating of balloons, but at the present day illuminating gas is so much cheaper, that the latter is generally used, although it is much heavier than hydrogen.

Diffusive Power of Hydrogen Gas.

It is not inappropriate to call attention here to certain interesting relations that hydrogen manifests towards gases and solids. Thus hydrogen possesses to a marked degree that curious facility of passing into and permeating other gases which is spoken of as its *diffusive power*. True, this power is possessed by all gases to a certain extent; but in rapidity of action none approach hydrogen. As early as 1825 a German chemist named Döbereiner announced his observations of this power. He noticed that upon collecting some hydrogen in a cracked jar, placed in a pneumatic trough, the hydrogen leaked out into the air more rapidly than the air went in. So that in fact the water of the trough rose on the inside of the jar. It has been since discovered that when almost any two gases whatsoever, if only of different densities, are separated by a partition having fine cracks or holes in it, the lighter gas always moves out into the heavier one more rapidly than the heavier gas moves in. As hydrogen is the lightest of all, of course it diffuses into other gases with the greatest rapidity.

In liquids, hydrogen does not ordinarily dissolve in any considerable quantity.

With solids however it displays some properties that are well nigh incredible. Thus it has a very curious aptitude for passing into the very interior of certain solid metals. The white, compact, solid metal palladium, although it has no visible pores, has the power of swallowing up into itself in some mysterious way nearly a thousand times its bulk of this gas; and again a thin sheet of this same solid metal, air-tight to all appearances, allows hydrogen to pass through it as easily as a sieve does water.

The Most Interesting Chemical Property of Hydrogen.

By all means the most interesting chemical property of hydrogen is its power to unite with oxygen. When it does so unite all the phenomena of combustion appear. These phenomena are generally the production of heat, light, flame, and the formation of some new chemical compound. So then when hydrogen unites with oxygen, it burns, it gives out light (although that light is of but feeble intensity), it gives out an enormous quantity of heat, it forms an oxidized product. This product is water, but water that—owing to the great heat of the combustion—is raised to the form of invisible vapor. When however a jet of hydrogen gas is burned under a bright but cool bell-glass, the deposit of mist quickly formed on the inside of the glass shows that the vapor produced by combustion has now condensed on the bell to minute liquid drops.

In the matter of the heat involved, hydrogen has the distinction of being above every other substance. One pound of hydrogen when burned under favorable conditions evolves heat enough to raise over *thirty-four thousand pounds* of water from zero centigrade to one degree centigrade, or nearly the same as from 32 degrees Fahrenheit to 34 degrees Fahrenheit. This expression of the calorific power of hydrogen has the same meaning as the following more technical one, namely : burning hydrogen affords over thirty-four thousand thermal units. Now

FIG. 9.—A glass tube held over a hydrogen flame, for the purpose of developing a musical note.

carbon, a fuel which nature has provided, and which is certainly admirably fitted to be man's chief combustible, yields but eight thousand thermal units of the kind just referred to, and for purposes of comparison it may be added that sulphur yields but two thousand thermal units.

Hydrogen Cannot Supply the Uses of Atmospheric Air.

Notwithstanding the remarkable evidences of chemical affinity suggested by what has just been said, hydrogen can in no sense act as a substitute for the atmospheric air. Thus it does not support animal life nor will it sustain the combustion of a candle. A living animal immersed in a room full of hydrogen would be drowned in it; a burning candle carried into such a chamber would be extinguished as if dipped in water. In fact the comparison with drowning is very proper, for in drowning a living animal the water does not chemically injure the organism; the hydrogen and the water, in the cases supposed, have similar action *in depriving both the animal and the taper of their requisite oxygen.*

The Uses to Which Hydrogen May be Put.

Hydrogen as the elementary gas finds but few applications in the arts. It is true that from what has been said, it appears as if its wonderful calorific power might be utilized in some of the arts where high temperatures are requisite. But the cost and difficulties attending its preparation, the liability to loss during its storage, and the danger from explosion while in actual use, these and other circumstances have led even the skilled artisan to content himself in most cases with other though inferior materials. But if the reader has attentively followed the introductory chapters of this work he must have perceived that hydrogen is made of great service in many of the measurements employed by the chemist. It has been noted that it is used as the standard of *equivalence or atom fixing power.* It

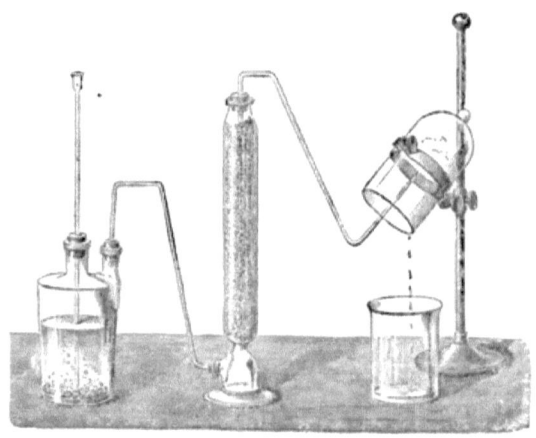

Fig. 10.—Disposition of apparatus for the production of water, by combustion of dry hydrogen in air.

has been spoken of as the standard of *atomic weight*, and from what has appeared in the remarks upon its lightness it will seem that it has been properly adopted as the *standard of density for gases*.

READING REFERENCES.

Cavendish, Henry
    Brougham, H.—Lives of Men of Letters and Science, etc. p. 429.
    Timbs, J.—English Eccentrics, etc. p., 132.
    Wilson, George.—Life of Cavendish. London. 1851.

Black, Joseph
    Brougham, H.—Lives of Men of Letters and Science, etc. London. 1845
  p. 324.

Diffusion of Gases.
    Graham, T.—Elements of Chemistry. 2 v. London. 1850. i, 84.
    ———Jour. of Chem. Soc. of London. xvii, 334.

Occlusion of Hydrogen by Palladium.
    Graham, T.—Jour. of Chem. Soc. of London. xxii, 419.

CHAPTER IX.

BALLOONS.

THE remarkable lightness of hydrogen early suggested the fitness of that gas for the inflation of balloons. From the earliest ages men have desired to navigate the air. The drudgery of land travelling over hills and mountains, over marshes and streams, through jungles and forests, has led men to prefer voyaging even by sea. Thus the people of the United States crossed the stormy Atlantic in large numbers long before they traversed the wilds of the American continent to the Pacific coast; and the early voyagers from New York to the Golden Gate of San Francisco preferred the water way, though it led them through an enormous distance and around the perilous Cape Horn, rather than undertake the shorter course over the Rocky Mountains. Even at a later date, the sea voyage to Panama, and across the Isthmus, and again by water way to San Francisco was the ordinary course until Pacific railroads created a land pathway from one side of the continent to the other. So men, envying the bird in its flight through the mobile air, have desired yet more to conquer its smooth courses, just as their keels have found a sliding pathway in the watery main. But no truly successful air-voyaging was possible until about one hundred years ago.

Invention of the Balloon.

In the year 1783 two brothers named Stephen Montgolfier and Joseph Montgolfier, succeeded in sending up into the atmosphere the first air-ship worthy of the name. They lived in France at a little town named Annonay, situated about forty miles south of Lyons, and at the junction of two small streams whose clear waters flow into the river Rhone. Here the brothers carried on with increasing skill and success the manufacture of paper, a business which their father had conducted there before them, and which in fact is carried on by their descendants of the same name even at the present

FIG. 11.—One of the balloons of the Montgolfier brothers.

day. The brothers, Stephen and Joseph, were skillful mechanics, and one of them, it is said, had studied Dr. Priestley's work on "Different Kinds of Air." This seems to have led him to to the idea of aerial navigation. However that may be, it is a matter of history that on the 5th of June, 1783, the two brothers sent up from Annonay a balloon about thirty-five feet in diameter. Naturally it was made of paper, though lined with linen. The ascensional power of this balloon was due to a proportional lightening of the air within it by the influence of heat. The heat was produced by the combustion of a large quantity of chopped straw, and also from burning wool previously saturated with a little alcohol. Probably the Montgolfier brothers did not then fully know why their balloon ascended: they appear to have thought that it arose because of the volumes of smoke that filled it. It is hardly probable that either Stephen or Joseph Montgolfier thought at that time of using hydrogen for their air-ship, notwithstanding its extraordinary lightness had been a matter of public scientific knowledge for six or seven years. This may seem the more strange in view of the admitted fact that as early as 1767 Dr. Black, of Edinburgh, had publicly demonstrated that a suitable vessel filled with hydrogen would ascend in the atmosphere as cork does in water. Of course they did not think of employing illuminating gas, because that substance was not then in public use.

The First Balloon Ascension in Paris.

The news of the wonderful and successful experiment at Annonay was quickly sent to Paris, where it produced a profound sensation. The interest extended from scientific men to the royal family and the court, and indeed to the entire population of the capital. For the French people— perhaps even more than other nations of Europe—seem to have been particularly interested at this time in the study of chemical and physical science. The king instantly issued a summons for the Montgolfiers to come to Paris. But the Parisians could not even await their arrival. The scientists of the capital, though but partially informed as to the character of the experiments performed at Annonay, at once set to work. They decided upon hydrogen gas as probably the best fitted for their purposes. Whereupon they filled a globular balloon with this gas, and prepared to try it in public upon the Champ-de-Mars. It is said that three hundred thousand people—that is, nearly half the population of Paris—gathered together, crowding every adjacent avenue, to witness the unparalleled undertaking. The liberation of the aerial messenger was announced to the public by a salvo of artillery. The balloon immediately shot upward and, piercing the clouds, was soon lost to view. When afterward it slowly descended it reached the ground some fifteen miles from Paris. Here a troop of peasants who detected the strange apparition, were at first struck with alarm but quickly rallied, attacked the strange monster and of course soon reduced it to shreds. The whole chain of circumstances created so much excitement that the Government thought proper to issue a proclamation upon the subject. A copy of this interesting document is here presented in its original form. Perhaps some readers will find the accompanying translation acceptable:

French Proclamation Respecting Balloons.

Avertissement au peuple sur l'enlèvement des ballons ou globes en l'air.

On a fait une découverte dont le gouvernement a jugé convenable de donner connaissance, afin de prévenir les terreurs qu'elle pourrait occasioner parmi le peuple. En calculant la différence de pesanteur entre l'air appelé inflammable et l'air de notre atmosphère, on a trouvé qu'un ballon rempli de cet air inflammable devait s'élever de lui-même dans le ciel jusqu'au moment où les deux airs seraient en équilibre, ce qui ne peut être qu'à une très grande hauteur. La première expérience a été faite à Annonay, en Vivarais, par les sieurs Montgolfier, inventeurs. Une globe de toile

Notice to the public relative to the ascension of balloons or globes into the air.

A discovery has been made to which the government considers it advisable to call public attention, with a view of preventing alarms which it otherwise might occasion among the people. Upon calculating the difference of weight between the gas called inflammable air and the air of our atmosphere, it has been discovered that a balloon filled with this inflammable air ought to rise of itself to a height in the sky such that the air within and that without will be in equilibrium, a condition which will not be reached except at a very great elevation. The first experiment of this sort has been made at Annonay, in Vivarais, by the Messrs. Montgolfier, the inventors.

et de papier de cent cinq pieds de circonférence, rempli d'air inflammable, s'éleva lui-même à une hauteur qu'on n'a pu calculer. La même expérience vient d'être renouvelée à Paris, le 27 août à cinq heures du soir, en présence d'un nombre infini de personnes. Un globe de taffetas enduit de gomme élastique, de trente-six pieds de tour, s'est élevé du Champ-de-Mars jusque dans les nues, où on l'a perdu de vue. On se propose de répéter cette expérience avec des globes beaucoup plus gros.

Chacun de ceux qui découvriront dans le ciel de pareils globes, qui présentent l'aspect de la lune obscurcie, doit donc être prévenir que, loin d'être un phénomène effrayant, ce n'est qu'une machine toujours composée de taffetas ou de toile légère recouverte de papier, qui ne peut causer aucun mal, et dont il est à présumer qu'on fera quelque jour des applications utiles aux besoins de la société.

Lu et apprové,
ce 3 septembre, 1783. DE SAUVIGNY.

A globe of cloth and paper one hundred and five feet in circumference and filled with inflammable air rose of itself to a height which the observer could not calculate. The same experiment has just been repeated at Paris on the 27th of August at 5 o'clock in the afternoon, in presence of a vast number of persons. A sphere of taffeta coated with gum elastic, thirty-six feet in circumference, ascended from the Champ-de-Mars even to the clouds, in which it became lost to sight. It is contemplated repeating this experiment with very much larger globes.

Anyone who discovers in the sky globes of this sort which present the appearance of the moon when slightly obscured, may therefore be warned that, far from being an alarming phenomenon, this is nothing but a machine always constructed of taffeta or of light cloth covered with paper, which cannot do any injury, and which it is thought will assume at some future time a form that will prove useful to the public.

Read and approved,
September 3, 1783. DE SAUVIGNY.

The enthusiasm created by the original experiment of the Montgolfier brothers led soon after to the election of both of them to the Academy of Sciences. Moreover their invention was not allowed to rest long in its original form.

As early as November of the same year, 1783, two French gentlemen had the courage to risk their lives in an ascension from Paris in a balloon of the Montgolfier construction. They floated freely away and made their landing in safety. One of them, however, De Rozier by name, on a later occasion attempted to cross the Channel in a double balloon, one part containing hydrogen, the other heated air in the Montgolfier style. But at a great altitude the hydrogen balloon took fire from the other, and De Rozier and his companion were dashed to pieces on the rocks of the French

Fig. 12.—Gay-Lussac and Biot making their balloon ascension for scientific observations in 1804.

coast. Since that early rash attempt thousands of interesting and safe balloon ascensions have been made, and increased knowledge of the scientific principles has largely contributed to the pleasure and comfort of the aeronaut. Yet the contrivance has been in most cases little more than a scientific toy.

The atmospheric air has thus far baffled the inventive power of man to such an extent that the balloon as a mechanical contrivance has been subjected to but few decided improvements since the Montgolfiers' first experiments, and ascensions have afforded comparatively meagre scientific or other results. Indeed the most of them have been conducted for personal gratification or popular entertainment.

Of course there are marked exceptions. Thus on the 24th of August, 1804, two of the youngest but most distinguished of French physicists, Messrs. Gay-Lussac and Biot, made an important ascension. Their voyage was upon the suggestion of the French Academy of Sciences, and they were well equipped with apparatus for making observations. Their results, particularly in magnetism, showed the same laws prevailing in the higher air as upon the earth. But as there were afterwards expressed some doubts as to the accuracy of these observations, Gay-Lussac made a later and higher ascent alone. On the 16th of September he attained an altitude of twenty-three thousand feet, the greatest reached up to that date. His experiments on this occasion verified those made before. Of particular interest was his test of the composition of the atmosphere. The bottle of air collected at this great height was found upon analysis to possess the same proportional amounts of oxygen and nitrogen as that collected at the surface of the earth.*

---

*Of Gay-Lussac and this ascension there is told a pretty tale, which I will not mar by making a translation:

"Parvenu à la hauteur de 7000 mètres, il voulut, dit-il, essayer de monter plus haut, et se débarrassa de tous les objets dont il pouvait rigoureusement se passer. Au nombre de ces objets figurait une chaise en bois blanc, que le hasard fit tomber sur un buisson, tout près d'une jeune fille qui gardait les moutons. Quel ne fut pas l'étonnement de la bergère!—Comme eût dit Florian.—Le ciel était pur, le ballon invisible.—Que penser de la chaise, si ce n'est qu'elle provenait du paradis?—On ne pouvait objecter à cette conjecture que la grossièreté du travail : les ouvriers, disaient les incrédules, ne pouvaient là-haut être si inhabiles. La dispute en était là, lorsque les journaux, en publiant toutes les particularités du voyage de Gay-Lussac, y mirent fin, en rangeant parmi les effets naturels ce qui jusqu' alors avait parut un miracle."—*Arago: Éloge de Gay-Lussac.*

The height of this ascent has since been surpassed by Messrs. Glaisher and Coxwell, of England, who on Sept. 5, 1862 attained an altitude of about thirty-seven thousand feet.

Recent Use of Balloons.

Hydrogen is the lightest substance known, and this consideration tends to make it a particularly favorable one for the inflation of balloons. But we have seen that it was not until after the Montgolfier experiments that hydrogen came into considerable use for this purposes. Hydrogen is still occasionally prepared for purposes of this sort. It is then produced by the action of sulphuric acid upon zinc. The equation already given explaining this action is as follows:

Zn	+	H_2SO_4	=	$ZnSO_4$	+	H_2
One atom of		One molecule of		One molecule of		One molecule of
Zinc,		Sulphuric acid,		Zinc sulphate,		Hydrogen,
65		98		161		2
parts by weight.		parts by weight.		parts by weight.		parts by weight.
163				163		

In case zinc is not at hand, iron-turnings have been made to answer the same purpose; and the chemical change in this event is represented by an equation of very similar form:

Fe	+	H_2SO_4	=	$FeSO_4$	+	H_2
One atom of		One molecule of		One molecule of		One molecule of
Iron,		Sulphuric acid,		Ferrous sulphate,		Hydrogen,
56		98		152		2
parts by weight.		parts by weight.		parts by weight.		parts by weight.
154				154		

Both of these methods of producing hydrogen are still somewhat used where balloons have to be inflated at points distant from a city gas supply.

But the manufacture of illuminating gas is now so general, even in small towns, that this substance is oftener used at the present day. The superior convenience with which it may be obtained makes it preferred to hydrogen, notwithstanding the greater ascensional power of the latter substance.

Balloons have been used somewhat in recent wars. Thus they were found of considerable service during the siege of Paris, particularly from September 23, 1870, to January 28, 1871. During these last four months of that siege sixty-two balloons left the city, and they carried out above two million letters and a great many homing pigeons. Some of the birds returned, escaping the Prussian sharpshooters, and brought with them letters and despatches, printed upon the thinnest of paper, in the form of microscopic photographs. The balloons also took out of Paris during the siege two especially notable passengers, the one, Leon Gambetta, head of the provisional government, who left the city for the purpose of conducting

FIG. 13.—Carrier pigeon having attached to his tail a quill containing microscopic photographs of despatches to be sent into Paris during the siege.

FIG. 14.—The tube of quill containing messages as attached to the tail-feathers of a carrier pigeon.

the public business in the provinces; the other, Prof. Janssen, who had the courage to venture out in the darkness of early morning so as to escape the rifles of the beleaguring forces. His voyage was for the purpose of reaching the station in Algeria from which he was to observe the total eclipse of the sun, to occur a few weeks later, December 22, 1870. Readers who are interested in the use of balloons during this memorable seige will find a most interesting account in Mr. Glaisher's book, *Travels in the Air*. It contains a description of the manufacture of air-ships in Paris, together with a list of the passengers and an account of the freight of those leaving the city when other means of communication with the outside world were cut off.

FIG. 15.—Owner's name on the wing of a pigeon.

FIG. 16.—Fac-simile of a microscopic despatch as sent by carrier pigeon. Letters and messages, public or private, to the number of about 5000, were printed on a large sheet of paper. Afterwards this sheet was reduced by photography to the size and appearance shown above.

The Centenary of Ballooning.

It is worthy of note that in August, 1883, the centenary of the experiment of the Montgolfier brothers was celebrated by their descendants and others at Annonay by a modern balloon ascension and other fêtes. These included the dedication of a monument to the two inventors. This monument is soon to be surmounted by a group in bronze representing the two brothers inflating their first balloon.

READING REFERENCES.

Balloons, Their Early History.

Figuier, L.—Les Aérostats et les Aéronautes. Revue des Deux Mondes. Oct. 1, 1850. p. 193.
[This interesting article will well repay the reader.]

Bierzy, H.—La Navigation Aérienne. Revue des Deux Mondes. Nov. 1863. p. 279.
[Claim is here made, in a general way, that the original invention of the balloon was made at the close of the 17th century by a Portuguese named Gusmao.]

Balloons, Their Recent Uses.

Glaisher, James, and others.—Travels in the Air. London. 1871.
Hofmann, A. W.—Chem. News. xxxii, 231, 241, 255, 265.

Balloons, Centenary of Their Invention.

London Graphic. Aug 25, 1883.

Balloons, Popular Account of

Harper's Magazine. ii., 168, 323; xxxix, 145.
Scribner's Monthly. i, 385.

Gay-Lussac.

Arago, D. F. J.—Oeuvres Completes. Paris. 1854-59. iii.
[The Boston Athenæum library has this work.]

CHAPTER X.

CHLORINE.

CHLORINE is a substance of very great commercial importance on account of its extensive use as a bleaching agent. Again it is a constituent of common salt, and in this form of combination it is both of great value as an article of food, and it is recognized as widely distributed. Thus it exists in salt, whether that substance is in the brine of the ocean or of mineral springs, or whether it occurs as a solid rock—as indeed it does in some parts of the world. At Wieliczka, in Austria, mines of solid salt have been worked for hundreds of years. So also at Cardona, in Spain, are what may be called quarries of this valuable mineral; while Cheshire, in England, furnishes immense solid deposits from which salt is obtained to supply the enormous industrial establishments using this substance for the production of chlorine and of compounds of sodium.

Chlorine was first recognized as a distinct substance, by a European chemist, Carl Wilhelm Scheele, known only to his neighbors as a humble apothecary. Scheele was born at Stralsund, a seaport town of Pomerania, situated on the little strait which leaves the island of Rügen in the Baltic Sea. He spent the principal portion of his life in Sweden, and on this account is often referred to as a Swedish chemist. Though living in great obscurity, he yet made many discoveries in chemistry which have rendered his name, otherwise almost unknown, one of the most brilliant in the annals of this science.

It is related of Scheele that the King of Sweden, Gustavus III., while on a journey outside of his own dominions, heard so much of the fame of this chemist, unknown to him before, that he regretted having previously done nothing for him. He therefore commanded that Scheele receive the honor of being created chevalier. "Scheele?" "Scheele?" said the minister charged with this duty. "This is very singular; what in the world has Scheele done?" The order was peremptory however, and Scheele was knighted. But, as the reader may perhaps divine, the honor designed for the acute discoverer fell upon another Scheele—not upon that Scheele unknown at court but illustrious among the scientists of Europe.

It was this obscure apothecary then, who added to the list of his other investigations a study of the properties of what was ordinarily considered a dull and uninteresting earthy substance called black magnesia. This study was repaid by the revelation of no less than four hitherto unknown substances: oxygen, barium, manganese, and finally chlorine. Scheele obtained the chlorine in the year 1774 exactly as it is done at the present day, namely, by bringing together the two substances, now called chlorohydric acid and black oxide of manganese, but then known as muriatic acid and black magnesia.

Scheele believed, and other celebrated chemists concurred in the opinion, that the greenish gas that he discovered was a compound substance. It was was not until thirty-six years later, that is 1810, that the distinguished English chemist, Sir Humphry Davy, demonstrated that this gas is not a compound, but is in fact a simple or elementary substance; and it was he who gave to it the name chlorine, a name derived from a Greek word ($\chi\lambda\omega\rho\acute{o}\varsigma$, chloros, meaning light green), conveying an obvious and convenient reminder of one striking property of the thing referred to.

How Chlorine is Obtained.

The preparation of chlorine is a very simple matter. It may be accomplished by placing some powdered black oxide of manganese, an

SIR HUMPHRY DAVY, Bart.:

Born in Penzance, England, Dec. 17, 1778; died in Geneva, Switzerland, May 29th, 1829.
"Davy, when not yet thirty-two years old, occupied, in the opinion of all those who could judge of such labors, the first rank among the chemists of this or any other age."

abundant mineral substance, in any deep glass vessel, and then adding to it four or five times its weight of chlorohydric acid. Anyone who performs the experiment will soon perceive the greenish gas rising higher and higher in the vessel, and will soon discover its choking and corrosive odor. Moreover the chlorine gas, which is two and a half times as heavy as air, accumulates within the flask and stays there some time. This is the process which has already been referred to as that which first revealed the gas to Scheele, and this process, with but slight modification, is that which to-day furnishes the enormous quantities of chlorine demanded by modern industries.

The Characteristics of Chlorine.

The three most striking properties of chlorine are its noticeable weight—greater than that of the air—its greenish color, and its exceedingly irritating ordor. Its influence on the animal organism is very violent: more than one example can be produced of fatal results following the inhalation of too large quantities of the gas. Thus Pelletier, a French chemist, died at Bayonne from the effects of inhaling a considerable quantity of chlorine, and Roe, a young Irish chemist of Dublin, lost his life from the same cause, while studying the properties of this gas.

Chlorine, as a *chemical agent*, manifests its activities in connection with two principal properties, namely: its affinity for hydrogen and its affinity for the metals. By this statement it is meant that chlorine manifests a strong tendency to combine with hydrogen, and to combine with metals, whenever these substances are accessible to it.

When it combines with hydrogen it forms the important compound designated by the formula HCl and called by the chemist chlorohydric acid, but known in commerce as muriatic acid.

When chlorine combines with the metals it forms chlorides of them. Thus with the metal sodium it forms the compound designated by the formula Na Cl and called by the chemist indifferently sodic chloride or chloride of sodium; these will be recognized as the chemical names for the important and well-known substance, common salt.

Chlorine and Hydrogen Combine.

Chlorine and hydrogen have a very strong tendency to combine with each other. They manifest this tendency in a variety of ways. Thus, if the two gases are prepared in a dark room, they may be there safely mixed together in a glass vessel; but if the sunlight is allowed to enter and fall upon the vessel, there is danger of its being shattered by the explosive violence with which the hydrogen and chlorine immediately unite. As a result of this combination, chlorohydric acid is produced.

The chemical change is represented by the following equation:

$$H_2 \;+\; Cl_2 \;=\; 2HCl$$

One molecule of Hydrogen, 2 parts by weight.	One molecule of Chlorine, 71 parts by weight.	Two molecules of Chlorohydric acid, 73 parts by weight.
73		73

Again, when chlorine is brought in contact with vegetable or animal substances, containing hydrogen, it proceeds to withdraw that hydrogen for its own benefit, even though these vegetable and animal compounds are thereby destroyed.

Although this operation, as well as the foregoing one, produces chlorohydric acid, yet neither method is suitable for a determinate preparation of that substance. It is usually better to prepare chlorohydric acid in another way. Thus it is easily produced by the action of sulphuric acid upon common salt.

Experimental Preparation of Chlorohydric Acid.

Any one who will take a little trouble may prepare chlorohydric acid in the way indicated.

The experiment should be conducted as follows:

Place a small amount of common salt (Na Cl) in a small retort; to it add enough concentrated sulphuric acid to make a thin paste; connect the neck of the retort with a clean test-tube containing a few drops of water. Now gently heat the retort; chlorohydric acid will be formed and will distil from the retort and condense in the receiver.

The chemical change is represented by the following equation:

NaCl	+	H_2SO_4	=	HCl	+	$HNaSO_4$
One molecule of Sodic chloride,		One molecule of Sulphuric acid,		One molecule of Chlorohydric acid		One molecule of Hydro-Sodic sulphate,
$58\frac{1}{2}$		98		$36\frac{1}{2}$		120
parts by weight.		parts by weight.		parts by weight.		parts by weight.

$156\frac{1}{2}$ $156\frac{1}{2}$

The product of the foregoing experiment may be tested in three ways and so shown to be in fact chlorohydric acid.

First: Take a minute drop on a glass rod and apply it to the tongue and observe the sour or acid taste.

Second: Take a drop on a glass rod and touch it upon blue litmus-paper. It should turn the paper red.

Third: Pour a few drops of the liquid into a solution of argentic nitrate (that is, nitrate of silver) in a test-tube or other convenient vessel: a white precipitate of argentic chloride will be formed.

The method of producing chlorohydric acid just described and illustrated, is followed in the manufacture of the substance for general chemical

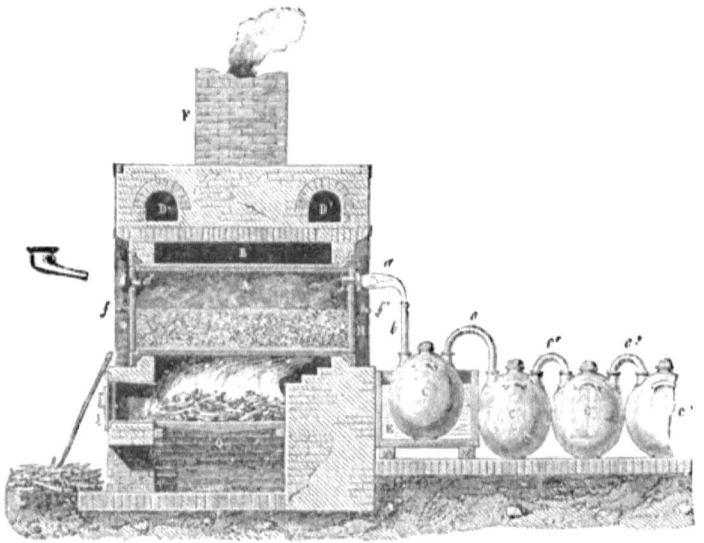

Fig. 19.—Section of furnace used for manufacture of chlorohydric acid. Common salt and sulphuric acid are placed in the large retort A; upon heating, chlorohydric acid passes into the receivers C, C, C.

purposes. It is also employed for the production of the enormous quantities of it incidentally used in the manufacture of bleaching-powder.

Experiments with Common Salt.

Chlorine has already been shown to combine with the metal silver producing the compound designated by the formula $Ag\, Cl$, and called argentic chloride and also chloride of silver. This substance may also be prepared very easily somewhat as follows:

Make a solution of nitrate of silver. Prepare it either by dissolving in water the crystals sold by apothecaries, or by dissolving a small piece of silver in nitric acid. Then make a second clear solution, by dissolving common salt in ordinary water. Add the salt solution cautiously, drop by drop to the silver solution. There immediately appear thick masses of white flakes which sooner or later fall to the bottom of the vessel. These flakes consist of the argentic chloride ($AgCl$), also called chloride of silver, already referred to.

The chemical change is represented by the following equation:

$$AgNO_3 + NaCl = AgCl + NaNO_3$$

One molecule of Argentic nitrate, $169\tfrac{1}{4}$ parts by weight.	One molecule of Sodic chloride, $58\tfrac{1}{2}$ parts by weight	One molecule of Argentic chloride, 143 parts by weight.	One molecule of Sodic nitrate, 85 parts by weight.

$$228 \qquad\qquad 228$$

This white precipitate produced in this experiment possesses some special interest from its use in photography. In fact chloride of silver, as a thin film upon the surface of the photographic paper, is the principal substance which, by its sensitiveness to light, produces the photographic picture. In fact any one, who tries the experiment last described, will soon observe, upon preserving the chloride of silver so produced, that it rapidly grows dark upon exposure to sunlight.

Bleaching-Powder.

The substance known as bleaching-powder may be spoken of in a general way as consisting of lime saturated with chlorine. This description points very justly to the method of producing the substance, but gives no idea of the chemical arrangement of the constituents. Scheele early noticed that chlorine gas possessed decided bleaching power, and the French chemist, Berthollet, soon called attention to the possible applications of the substance in the bleaching industries. But its annoying odor made it impracticable to use chlorine on any large scale in the state of gas, and forbade the use of it even when dissolved in water. At length, twenty years after the discovery of the gas—that is in 1798—the plan of absorbing chlorine in lime was hit upon, and here may be discovered the beginnings of the bleaching-powder industry, now one branch of the alkali trade, the greatest chemical industry conducted by man. This bleaching-powder, at first a mere chemical curiosity, is now manufactured by the thousands of tons, and is used in the bleaching of cotton and linen goods, both in the form of cloth and in the form of the various kinds of paper.

In another place reference is made to the vast proportions attained by the alkali industry, meaning the manufacture of certain compounds of sodium, the one produced in largest quantities being doubtless sodic carbonate (Na_2CO_3), commonly called soda-ash. In trade this substance is called an alkali because of certain alkaline properties it possesses, but more strictly speaking it is called a salt—sometimes an alkaline salt. In chemistry the single term alkali is reserved for certain compounds called hydrates, of which indeed sodic hydrate—having the formula $NaOH$, and often called caustic soda—is an appropriate example. This latter compound is at present manufactured on a large scale in connection with soda-ash. Now although the Leblanc process—which has long been used for manufacturing soda-ash—is at present meeting with a powerful and successful rival, yet the older process has still a strong hold upon life in the fact that it gives rise, as a convenient incidental product, to vast quantities of chlorohydric acid. The meaning will be better understood when it is explained that the first step of the Leblanc process is to add sulphuric acid to common salt. Two substances are here produced: the one is sodic sulphate, to be carried forward until it is turned into sodic carbonate; the other substance is chlorohydric acid, a compound largely used in the arts, and especially in the manufacture of bleaching-powder.

Fig. 21.—Bleaching of pulp for manufacture of paper.

In the production of bleaching-powder, the first step is to mingle this chlorohydric acid and manganese di-oxide. Chlorine gas is thus generated, much as it is when the experiment is conducted on a small scale as already described. The chlorine so generated is passed into a chamber provided with shelves and containing slaked lime. Hereupon the lime absorbs the chlorine, giving rise to a new substance called bleaching-powder—also known as chloride of lime. From what has been said it is evident that chemists

Fig. 22.—Apparatus for producing bleaching-powder (by passing chlorine gas, generated in A, upon quick-lime spread upon the shelves).

know perfectly well what elementary substances enter into this compound. But there are decided differences of opinion as to the exact way in which the atoms are arranged. Bleaching-powder is generally considered to be a chemical union of calcic hypochlorite and calcic chloride with the addition of calcic hydrate. The following representation may serve as a formula for the compound:

$$CaCl_2O_2 \quad + \quad CaCl_2 \quad + \quad CaO_2H_2$$
(Calcic hypochlorite.) (Calcic chloride.) (Calcic hydrate.)

FIG 20.—Representation of the old method of bleaching cotton and linen goods on lawns.

The use of bleaching-powder offers certain advantages. The following are some of them:

Fig. 23.—Apparatus for "souring" cotton cloth by passing it into dilute acid, before submitting it to the action of bleaching-powder.

—The compound is itself white.
—It is a powder which can be easily handled, packed and transported.
—With reasonable precautions, the active bleaching agent chlorine

is retained by the powder in available form for a considerable length of time.

— In actual use in the process of bleaching, the entire amount of chlorine originally stored up in the powder may be liberated in contact with the goods to be bleached.

— The liberation of this chlorine is easily affected. The addition of almost any acid will accomplish it : even the carbon dioxide of the atmosphere will suffice.

In the bleaching of cotton goods chlorine is not the only agent relied upon, though it seems to be an essential one. At least three other substances are employed to contribute to the bleaching. Each of them either removes some colors or stains from the goods, or so modifies them that the solution of bleaching-powder — one of the last agents to be employed — can the easier finish its work. The three substances referred to are milk of lime, diluted sulphuric acid, and sodic carbonate, also called soda-ash.

The pieces of cloth, being sewed together into continuous strips many miles in length, pass from one liquor to another, with washings in water at proper times, until finally, after being fully whitened by the chlorine preparation and then receiving the final washing in water, they emerge from the works, completely bleached.

READING REFERENCES.

Alkali Trade, in its Various Branches.

  Claus, C.—Chem. News. xxxviii, 263. (Ammonia soda.)
  Davis, G. E.—Chem. News. xxxii, 164, 174, 187, 198, 210, 238.
  Hargreaves, J.—Chem. News. xlii, 322.
  Kingzett, Charles T.—The Alkali Trade. London, 1877.
  Lunge, G.—Jour. of Chem. Soc. of London. xliv, 524, 528.
  Mactear, J.—Chem. News. xxxv, 4, 14, 17, 23, 35; xxxvii, 16.
  Schmidt, T.—Chem News. xxxviii, 203. (Ammonia soda.)
  Weldon, W.—Chem. News. xlvii, 67, 79, 87. (Present condition of soda industry.)

Bleaching Powder.
>Jurisch, K.—Jour. of Chem. Soc. of London. xxxi, 350.
>Kingzett, C. T.—*loc. cit.* xxviii, 404.
>Kopfer, F.—*loc. cit.* xxviii, 713.
>Lunge, G.—Chem. News. xliii, 1.
>Stahlschmidt, C.—Jour. of Chem. Soc. of London. xxxi, 279.
>Wolters, W.—*loc.cit.* xxviii, 404.

Chlorine Industry, Future of
>Hurter, F.—Jour. of Chem. Soc. of London. xlvi, 225.

Chlorine, Preparation of
>Berthelot.—Annales de Chimie et de Physique. 5 Sér. xxii, 464.

Davy, Sir Humphry.
>Davy, John.—Collected Works and Memoirs of Sir H. Davy. London, 1839.
>Paris, John A.—Life of Sir Humphrey Davy. London, 1831.
>Brougham, H.—Lives of Men of Letters, etc. London, 1845. p. 448.
>Cooke, J. P.—Scientific Culture. Boston, 1881. p. 11.

Salt Mines of Europe.
>Harper's Magazine. i, 759.

Scheele, C. W.
>Hoefer, F.—Histoire de la Physique et de la Chimie. Paris, 1872. p. 497.

CHAPTER XI.

BROMINE.

BROMINE is an elementary substance, which was first recognized as such in the year 1826. It was detected by Antoine Jerome Balard, a French chemist, who, at the age of twenty-four, was so fortunate and skillful as to discover this interesting substance. He lived at Montpellier, not far from Marseilles and but a few miles from the Mediterranean. The waters of this great inland sea contain about one-tenth more mineral salts than those of the larger oceans, and so it has long been the custom along the southern coasts of France to evaporate these waters for the production of common salt. After this principal constituent is removed, there remains a strong brine called bittern. While experimenting upon this bittern, Balard was struck by a peculiar orange-red coloration of great intensity which appeared at certain stages of his work. Upon further study, he was able to demonstrate that this color was due to an elementary substance hitherto unrecognized. Thus he had the felicity of securing for his name permanent renown as one of the few philosophers who have been able to detect a new member of that family of prime and fundamental materials from which is built the structure of the universe.

It has already been stated that the elements at present acknowledged are less than seventy in number, and some of these were known to the ancients. In some cases, a single individual has been able to recognize several new ones: thus Scheele has already been mentioned as the discoverer of manganese, barium, and chlorine. So it appears that of all the eminent men, whose conscientious labors contribute to the building of the science of chemistry as a noble and harmonious edifice, necessarily but few can possibly hope to attain the specially conspicuous honor of having their names forever associated with the first discovery of any of the primary elements. An interesting story is told of the eminent German chemist, Justus von Liebig, in connection with this particular subject. Some years before Balard's discovery there was sent to Liebig, from a German establishment where salt brines were employed, a flask—whose contents were afterwards found to contain bromine, or at least to be very rich in bromine —with the request that he examine the contents. The general appearance of the substance seemed to be that of chloride of iodine, and this circumstance led Liebig to neglect making a more searching investigation. After Balard had published his discovery, Liebig perceived his own unfortunate oversight, and occasionally, of course not without some bitter regret, he displayed to his friends this interesting flask, to show them how one might fail to make a discovery of the first importance by reason of some trifling oversight.*

Distribution of Bromine.

The name bromine is derived from a Greek word ($\beta\rho\tilde{\omega}\mu o\varsigma$, *bromos*, a bad smell) which suggests the very pungent odor of its vapor. The substance occurs in the brine of the ocean and in that of mineral springs. But of course it does not exist there in the uncombined form; instead it is united with certain metals in the form of bromides. In sea-water the principal bromide is bromide of magnesium ($MgBr_2$).

---

*SCHUTZENBERGER, PAUL: *Traité de Chimie Générale*. Paris, 1880. i, 375.

Experimental Preparation of Bromine.

Bromine may be prepared by anyone who is willing to take a little trouble.

Place in any suitable glass vessel a small amount of manganese dioxide, some potassic bromide (commonly known as bromide of potassium), then some water, and finally a small quantity of chlorohydric acid. Bromine is almost instantly liberated, and shows its presence by imparting to the liquid an orange hue. If the vessel is covered lightly, and then gentle heat is applied to it, the bromine will be expelled from the liquid and will appear above it as a heavy vapor of a rich reddish-brown color. Some care must be exercised however in conducting this experiment, since the vapor is very irritating to the eyes and also to the throat, and it has a general corrosive effect upon most substances with which it comes in contact.

Chemical Properties of Bromine.

In its chemical relations bromine shows very decided resemblances to chlorine, having affinities for the same substances, only less in intensity. Since its discovery it has found a considerable number of uses. Thus it is an important substance in the processes of photography; and the enormous expansion and growth of this art within a very few years has required in the aggregate large quantities of bromine. The considerable demand for bromine, which at first increased its price, has produced, as might have been anticipated, a stimulating influence upon the manufacture of it. This has led to greatly increased production of the substance, not only in Europe but also in the United States. In Pennsylvania, Ohio, and West Virginia it has become an important article of manufacture; in fact, the United States now furnishes the largest proportion of the entire amount of the material produced in the world.

One of the most important compounds of bromine is that produced by its union with silver. We refer to argentic bromide (commonly called bromide of silver, Ag Br). This substance may be easily produced by the following simple experiment.

FIG. 24.—Louis Jacques Mandé Daguerre, from whom the daguerreotype was named: born at Cormeilles, (France,) 1789; died, 1851.

To a solution of potassic bromide in water add a water solution of argentic nitrate; a white, or yellowish-white precipitate immediately appears.

The chemical change is represented by the following equation:

$$KBr + AgNO_3 = AgBr + KNO_3$$

One molecule of	One molecule of	One molecule of	One molecule of
Potassic bromide,	Argentic nitrate,	Argentic bromide,	Potassic nitrate,
119	169¾	187½	101
parts by weight.	parts by weight.	parts by weight.	parts by weight.

$288\frac{1}{2}$ $288\frac{1}{2}$

This argentic bromide produced, at first nearly white in color, has the power of becoming black upon exposure to light, and it is this important property which makes the substance suitable for use in the processess of photography.

Again in the form of potassic bromide, bromine has had a very wide and beneficent use as a remedial agent; it is still largely used in the manufacture of the salt mentioned.

READING REFERENCE.

Liebig, His Life-Work in Chemistry.
    Hofman, A. W.—Jour. of Chem. Soc. of London. xxviii, 1065.

FIG. 25.—Photographer at work in a room lighted through a window of red glass. (Red glass cuts off the chief actinic, or chemical, rays of sunlight.)

CHAPTER XII.

IODINE.

IODINE clongs to what may be called a chemical family, the other members being chlorine and bromine. All three of these elements are found in sea-water, but in very different quantities. Thus chlorine is extremely abundant; bromine is in the water in minute quantities, while iodine exists there in amounts that are exceedingly small. They all exist as salts, of which of course chloride of sodium is by far the most abundant. It has already been shown that bromine is obtained from sea-water, after enormous amounts of the water have been concentrated by evaporation. But iodine, the third element of this group, exists in sea-water in quantities so very minute that it cannot be extracted from it at any practicable cost. Even the concentration method, just alluded to, is not applicable in the case of iodine. It happens however that sea-weeds have the power of extracting from sea-water even the exceedingly minute amount of iodine, or of iodides, that the water contains; and moreover when sea-weeds are burned, iodides are found in their ashes.

The Discovery of Iodine.

The discovery of iodine is associated with the history of certain of the most important and interesting products of the chemical arts. It also has a striking connection with some of the political and military affairs in France, and indeed in Europe, in the early years of the present century. Finally, its great usefulness to mankind is in marked contrast with the misfortunes that overtook its discoverer.

The discovery of iodine is directly referable to the old soda industry. The term soda is a general one, and it was formerly used to include several different chemical compounds manufactured from the ashes of sea-weed. Decidedly the most important of these is sodic carbonate. This substance

FIG. 26.—Gathering the harvest of sea-weed for the manufacture of soda-ash.

has a well marked alkaline reaction, and although not an alkali in the strictest chemical sense, it is yet the principal product of that greatest of all the chemical industries known as the alkali trade. (See pp. 94 and 99.) During the last sixty years, and after many early trials and failures, the production of the various alkaline compounds of sodium has risen to enormous proportions, such that in England alone the daily product of sodic carbonate, the principal one, is probably more than two thousand tons. This vast amount of alkali is consumed by civilized peoples in some of their most extensive industries such as the manufacture of soap and of glass, and in many processes of bleaching. The extension of these branches of business has of course gone hand in hand with the increased production of alkali. Indeed, on the one side there has been a steady diminution in price, and on the other a steady increase in consumption; probably each circumstance may be considered as both cause and effect of the other. Prior to 1793 however, the demands for alkali—vastly smaller than to-day—were all satisfied by the material obtained from the ashes of marine plants. Thus along the coasts of Great Britain, France, and especially of Spain, sea-weed of various kinds was gath-

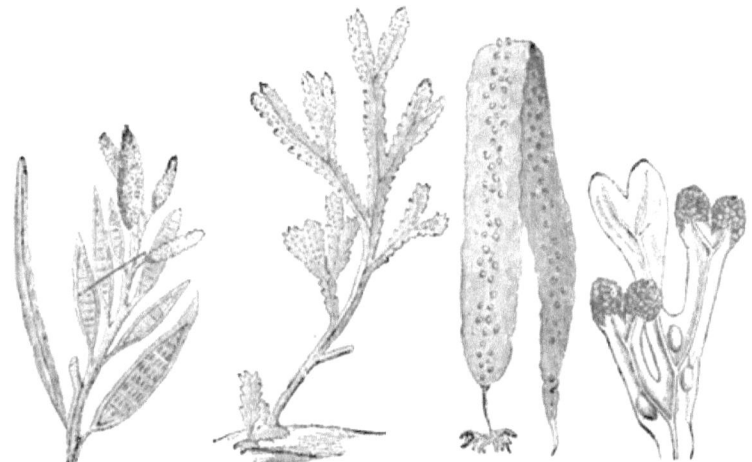

FIG. 27.—Varieties of sea-weed used to produce varech.

ered as a very important harvest. Some of the weed was used as a fertilizer of the soil; more was dried and burned for the sake of the ashes. On the British coast the ash was known as *kelp;* that produced on the coasts of Normandy was called *varech*; and that produced on the Spanish coast went by the name of *barilla*.

Now one of the important indirect effects of the French Revolution was that felt by the consumers of the old-fashioned alkali. In 1793 an embargo was put upon the supply of alkaline ashes, such as kelp and barilla, into France. But the French demand for alkali, not only for ordinary purposes, but also for the production of the great amounts of saltpetre required for the manufacture of gunpowder, was imperious. The immediate effect, therefore, was to create the sudden development of a process called the Leblanc method, by which alkaline compounds of sodium are made from common salt. Notwithstanding the stimulus of the prohibitory embargo and the fostering help of the government of Napoleon Bonaparte, the complexity of the Leblanc process was such that it was slow in gaining a foothold as a practical industrial method. But after its first successful establishment as a regular business and up to almost the present day the application of this process has continually widened, and the method has held undivided sway in its important field. In the year 1811 Bernard Courtois, a French chemist, was engaged, just as other manufacturers were, in the production of nitrate of potash, or saltpetre, for use in gunpowder. In the course of this work he employed soda obtained from varech. In order to separate the alkali from the varech in a more refined condition the raw varech was subjected to a very careful purification. At certain stages of his experiments Courtois discovered that the addition of sulphuric acid gave rise to the production of a magnificent violet vapor. He did not make the matter public however until late in the year 1813, when he brought the subject to the attention of Sir Humphry Davy, the distinguished English chemist, who was then visiting Paris. The next year, 1814, the substance was carefully investigated by Gay-Lussac, who gave to the world a very full description of its properties, and who called it iodine from a Greek word ($ιωειδης$, *ioeides*, violet colored), suggesting the striking and characteristic color of its vapor. The political events of 1815 ruined the business of Courtois, and he sunk into poverty from which he was not able to recover, until finally he died in 1838, poor and almost forgotten, leaving a widow who was forced to win her bread by the labor of her hands.

Present Sources of Iodine.

Although kelp, varech and barilla are no longer used for the direct purpose of affording alkali, they are still produced with a view to their

IODINE.

yielding iodine. On the rough and stormy coasts of Scotland, Ireland, France and Spain, large quantities of sea-weeds are cast ashore. They are collected, they are dried in the sun, they are then burned, and their ashes are employed—but principally in the manufacture of iodine. Thus on the coasts of Brittany and Normandy the occupation of collecting weeds occupies three or four thousand families for the larger part of the year.

Experimental Method of Preparing Iodine.

Iodine may be prepared in a manner closely resembling the process already described for bromine; that is, by placing in a suitable glass vessel a small amount of manganese dioxide, some potassic iodide (commonly known as iodide of potassium), then some water, and finally a small quantity of chlorohydric acid. Iodine is almost instantly liberated, and shows its presence by imparting to the liquid a brownish color. If the vessel is covered lightly and then gentle heat is applied to it the iodine will be expelled and appear in the vessel above the liquid as a heavy vapor of a rich violet color. This vapor readily condenses on the upper and colder portions of the vessel in the form of minute crystals of a color almost black. This is almost precisely the method employed on the large scale for the production of iodine from kelp.

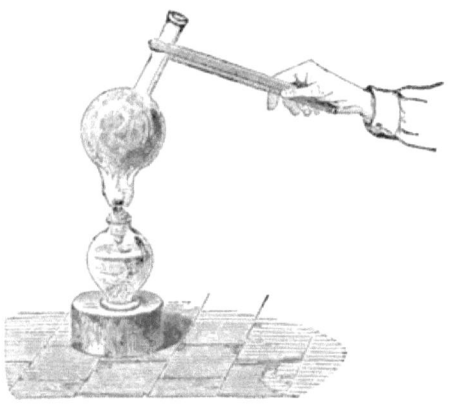

FIG. 28.—Changing iodine to a violet vapor by means of heat.

Chemical Properties of Iodine.

The chemical characteristics of iodine are throughout closely allied to those of chlorine and of bromine, only in general iodine may be said to have weaker chemical affinities than either of the other two.

Iodine produces compounds of the same general type as the others, and of this an example is found in argentic iodide. The following method of pro-

Fig. 29.—Apparatus used in the manufacturing process for obtaining iodine. The retorts C, C, are surrounded by sand (sandbath); the heat drives iodine, in form of vapor, into the receivers, A, A, where it solidifies.

ducing it can be followed by almost any one. Prepare a solution of nitrate of silver in water, and then add a water solution of potassic iodide; a chemical change takes place, with the production of a yellowish-white precipitate. This precipitate is argentic iodide. Upon exposure to sunlight it readily changes in color, becoming almost black. This is an important characteristic and is made use of, as is the same property possessed by argentic bromide and also by argentic chloride, in the production of the photograph. And while it is a fact, and one well known, that many of the salts of silver blacken more or less upon exposure to sunlight, it is found that the chloride, the bromide, and the iodide, have properties particularly fitting them for the purposes of photography. In discussing bromine, reference was made to the influence of the great expansion of the photographic business; and this circumstance has stimulated the demand for iodine just as for bromine. It was also pointed out, that potassic bromide is an important remedial agent; potassic iodide is likewise of great medicinal value.

Starch as a Test for Iodine.

Iodine, when in the free or uncombined condition, has a remarkable and very peculiar way of attaching itself to granules of starch.

This property may be demonstrated by a simple and attractive experiment. Thus if starch is boiled with water and then the hot mass is poured into cold water, minute particles of starch distribute themselves through the liquid. If to this liquid a very small amount of free iodine, in the form of a solution, is added, the starch instantly takes on a deep blue color. If to another portion of the same or similar starch suspended in water, iodine is added *in a combined form*—that is as potassic iodide for example—absolutely no change of color is detected. These two experiments show that the iodine only attacks starch when the iodine is *free and uncombined*.

READING REFERENCES.

Chlorine, Bromine, Iodine, and Fluorine.
    Myllus, E.—Chem. News. xxxiii, 244, 253; xxxiv, 5, 13, 25, 33, 45, 55, 66, 78, 86, 118, 139, 149, 166, 180, 188, 197, 215, 233.

Iodine, Manufacture of
    Schmidt, T.—Chem. News. xxxvii, 56.
    Stanford, E. C. C.—*Loc. cit.* xxxv, 172.

CHAPTER XIII.

FLUORINE.

F Fluorine it is necessary to made the remarkable statement that it has never been known to be produced isolated, that is in a separate or uncombined form. Many experiments have been performed for the purpose of reaching this result, and though none of these have resulted in the production of the element sought, they lead us to believe that the element is a gas. For, if it were a solid or a liquid at ordinary temperatures, it may be safely supposed that some processes that have been devised would be capable of producing at least a small quantity of the elementary substance, and that from this the observer would be enabled to recognize and discover at least some of the properties of the fluorine itself.

Properties of Fluorine.

There are a number of compounds known whose various properties, powers of chemical interchange, and special molecular weights, clearly point out the existence in them all of a certain peculiar element analogous in many respects to chlorine, bromine, and iodine. To this element the name fluorine has been given. Although, as before intimated, it has not been known to have been obtained *liberated*, its properties in these combined forms have been carefully studied and well made out. Thus, like chlorine and its family associates it combines with hydrogen to form an acid, fluohydric acid (HFl), properly comparable with the acids formed by the three elements last discussed:

Chlorohydric acid,	HCl.
Bromohydric acid,	HBr.
Iodohydric acid,	H I.

It also combines with the metals to form fluorides. The best example of these fluorides is that compound in which fluorine most commonly occurs in nature: that is fluor-spar, the mineral substance whose chemical name is calcic fluoride, and whose composition is expressed by the formula $CaFl_2$.

The property above all others that is characteristic of fluorine is, however, its striking affinity for silicon. With this element it readily combines under almost any circumstances. More wonderful still, the compound produced with it is a gas. Now in general the compounds of silicon are solids. These solids are many of them familiarly known in those materials which constitute the principal portions of the stable earth on which we tread, of the rock beneath it and of the enduring mountain masses that here and there pierce through the soil and raise their crests above the general level. The majority of these earthy and rocky substances are silicates. It is apparent then that the compounds of silicon are types of solidity and stability. They cannot be melted except in the most powerful heating appliances, and the chemist can hardly imagine conditions such as would change them into vapor. So then it seems strange and almost contradictory that fluorine should have the power of attacking compounds that seem to be the embodiments of permanency itself;—yet it readily does so. Thus if fluohydric acid comes in contact with silicon, whether that substance is in combination as sand or as hard rocky minerals, the fluorine atoms pluck out the silicon and then they fly away together in the form of gas or vapor. Again, fluohydric acid may be spoken of as the unique agent that readily attacks glass and dissolves, and even destroys, this ordinarily unchangeable substance.

Finally, there may be added what can be said of no other element, namely: that fluorine is never known to form any compound with oxygen.

Discovery of Fluohydric Acid.

It is not easy to refer the first knowledge of fluorine to any particular discoverer. Perhaps however renewed mention of the ingenious Scheele is not out of place here; for it seems to have been he who for the first time, and as early as 1771, recognized fluohydric acid as a special acid.

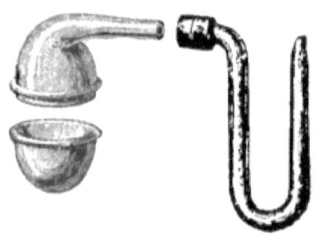

Fig. 30.—Platinum retort and receiver shown with its several parts separated.

He called it fluoric acid, but he did not obtain a correct idea of its composition. Scheele prepared the acid from a well-known mineral, fluorspar, and by the addition of sulphuric acid. This operation cannot be performed to advantage in a glass or porcelain vessel for they contain silicon, and as has been suggested already, silicious matters are freely attacked by the acid produced. The decomposition therefore is commonly conducted in a retort of lead, or in one of platinum, and the acid produced is collected in a receiver, also constructed of one of these metals.

The chemical change is represented by the following equation:

$$CaFl_2 + H_2SO_4 = 2HFl + CaSO_4$$

One molecule of Calcic fluoride,	One molecule of Sulphuric acid,	Two molecules of Fluohydric acid,	One molecule of Calcic sulphate,
78	98	40	136
parts by weight.	parts by weight.	parts by weight.	parts by weight.
176		176	

Ordinarily the product is a liquid, and consists of water holding in solution the fluohydric acid (HFl). It is possible however to prepare the acid free from water, and still in a liquid form. But in this condition it is one of the most dangerous, poisonous, and corrosive substances known. It produces painful burns if it falls upon the flesh, and fatal results have been known to follow injuries received from it. Thus in 1869, Professor Nicklès, an eminent French chemist, died from injuries sustained by the accidental inhalation of fluohydric acid vapor, while studying the properties of the substance.

Etching Glass by Fluohydric Acid.

The effect of fluohydric acid upon glass may be shown in attractive form, and without much difficulty or danger, by the help of a small dish of lead and a plate of glass to cover it. These being provided, the experiment may be conducted somewhat as follows: Melt a little beeswax upon the glass so that the wax may form a thin film upon one side of it. Then allow the wax to cool and harden. Next, by use of any convenient pointed instrument, draw some sketch or design deep in the wax—in fact, to the surface of the glass. Next place some powdered fluor-spar in the leaden dish, and add to it some concentrated sulphuric acid. Now cover the dish, with the glass already prepared, in such a way that the sketch or design is turned downward so as to receive the fumes of fluohydric acid as they rise from the mixture in the dish. It is easily understood from what has been said already that the fluohydric acid will attack the glass, carrying away some of its silicon in the form of gas or vapor. As a result of this action, minute channels are formed in the glass. When the experiment is thought to be sufficiently advanced, the wax may be removed

FIG. 31.—Platinum retort and receiver shown as arranged for production of fluohydric acid (HFl).

from the plate by melting it off or otherwise; thereupon it will be discovered that the glass has actually become etched or engraved by the fluohydric acid gas.

In 1788 Puymaurin presented to the French Academy of Sciences such a glass plate, upon which there was a beautiful fluoric etching representing Chemistry and Genius weeping at the tomb of Scheele, who had contributed so much to the history of fluohydric acid. "This work," says Haüy, "was of interest to the Academy on account of the fitness of the subject as well as the elegance of its execution."

Practical Application of Fluohydric Acid.

Fluohydric acid, formerly a mere chemical curiosity, has now become a familiar article upon the shelves of the druggists. It is sold in gutta-percha bottles with rubber stoppers. It is often used by jewelers to correct errors in the application of silicious enamels upon their work. Thus if the enamel has been incorrectly placed, it may be removed by fluohydric acid and afterward a new portion may be introduced in the proper position. Again, it is largely used in the decoration of artistic glass

FIG. 32.—Leaden tray and glass plate. The tray is intended to receive the materials for production of fluohydric acid; the plate is represented as covered with a varnish, through which a sketch has been drawn, preparatory to etching.

objects, such as globes for gas chandeliers, and the multitude of articles of table glass ware. In engraving such objects, they are first covered with a suitable varnish that will resist the fluohydric acid, then the design is drawn through the varnish with a sharp needle; afterward the article is exposed to the gas and etched in a manner similar to that already described.

READING REFERENCES.

Fluohydric Acid.

    Gore, G.—Jour. of Chem. Soc. of London. xxii, 368.

Fluorides.

    Fremy, E.—Annales de Chimie et de Physique. 3 Sér. xlvii, 5.

CHAPTER XIV.

OXYGEN.

OXYGEN may justly claim a high degree of importance as a subject for the study alike of the professional chemist and the casual reader. This importance depends upon a variety of considerations. Among them are the surpassing abundance of the substance itself, the great number of compounds into which it enters, the activity of its chemical powers, and finally, the interesting circumstances under which its distinct recognition, or, as perhaps we may say, its discovery, was attained.

Its great abundance has been pointed out already in the declaration that oxygen makes up, by weight, fully one-half of our terrestrial globe—including earth, ocean and air. The air is about one-fifth oxygen by weight; all water, wherever existing, is sixteen-eighteenths oxygen by weight, while quartz, sand, and other similar widespread and most commonly occurring mineral matters, are a little more than one-half oxygen. Other solid matters than the rocks, such as most parts of the material structures of animal and vegetable beings, contain oxygen as an important constituent element. While thus we have scanned the great multitude of substances spread immediately about us by the hand of nature, and found oxygen in them all, it is none the less true that oxygen is an important factor in artificial products—that is, those resulting from man's manufacturing operations.

Chemical Activity of Oxygen.

Again, oxygen plays a part of exceeding activity in some of the grandest chemical processes of nature and of the arts.

For example, it is essential to the vital processes of all animals. Wherever a living being inhales the breath of life, whether from the fresh air of the mountain tops, or from the populous streets of the swarming metropolis, or from the solitary deck of the bark that creeps with the ocean's currents; or wherever the humbler servants of man's table find their way through unexplored depths of the ocean and pluck from its waves the modicum of life-giving gas dissolved within them; there is this wonderful agent, which has no substitute, sustaining by active processes truly chemical, that vitality of man or of beast which gives to nature its forms of highest beauty and most admirable intelligence.

Again, oxygen is the necessary agent in all ordinary combustions. So wherever a faggot, glowing beneficently in a sparsely peopled forest, helps to sustain man's vital spark; or, where in a highly civilized community, the fires on the altars of modern industry draw from the flinty rocks the metals that serve to give employment to millions of children of toil;—there oxygen is ever active, the true supporter of the combustion of all those flames which in the past have served as signs of life and civilized activity, and which are still the best symbols of vitality and intelligence.

The Discovery of Oxygen.

The first discovery of oxygen is usually attributed to Dr. Joseph Priestley, an English clergyman and student of natural science. He lived in a time when men's minds all over Europe were strongly drawn toward the pursuit of chemical knowledge. In fact, at almost the same moment that Priestley was enthusiastically conducting his experiments, Scheele was also producing oxygen in his apothecary's chamber in Sweden. And the brilliant Lavoisier, prominent among the men of distinction who thronged the gay capital of France, was also working in the same direction; it was he, who said about oxygen in one of his own chemical

JOSEPH PRIESTLEY,
Born near Leeds, England, March 13, 1733; died in Northumberland, Pa., February 6, 1804.

works: "Cet air que nous avons decouvert presque en même temps, Dr. Priestley, M. Scheele et moi," so that he is sometimes declared by his enthusiastic countrymen to be entitled to the merit of the earliest discovery of this most magnificent of elements.

Priestley's life included ample materials for a romance. On the one hand, the ingenious discoverer in physics and chemistry and the friend of that Benjamin Franklin—who was then minister at the brilliant court of France from a handful of colonies that appeared capable of being plucked up by the roots, but were instead destined to grow to an unrivalled empire—himself a figure in a romance; and, on the other side, a preacher to a dissenting congregation; a victim of public odium for his liberal opinions on religious and political subjects; his house set on fire by a mob, his apparatus wrecked, his library cast to the winds; finally, an emigrant with his wife and children to an almost unknown village in Pennsylvania, whose little burial-ground still gives his bones repose;—these are but brief suggestions of the trials of this perturbed spirit, in his life "sadly driven about and tossed," now cherished as one of those who in the realm of thought has made no mean contribution to the glory of the English name.

Dr. Priestley prepared oxygen from red precipitate of mercury, a substance now designated by the name mercuric oxide and by the formula HgO. Heating this substance in a receiver and by means of a burning glass or lens, he observed that a peculiar kind of air was evolved. He further discovered that this air had an unusually stimulating influence upon burning bodies, and was well suited for the respiration of living animals. Priestley's prime experiment was performed on the first day of August, 1774, a date which may be accepted as almost the birthday of modern chemistry.

Like many other great discoverers, Priestley was, to a certain degree, anticipated. Thus a certain John Mayow, an English physician, fully a hundred years before the time of Priestley's experiment, enunciated the doctrine that the atmosphere contains an air, in a certain sense the essential food of animal life and of flame. But these wonderful views of Mayow, brought forward too early for the state of thought at his time, lay dormant and unproductive for an entire century.

First Method of Preparing Oxygen.

Oxygen may be prepared in many ways, but only two need receive attention here. The first method is Priestley's. If the red oxide of mercury is heated over a powerful gas flame and in a tube of not easily fusible glass, the oxygen passes from the metal and may be carried by any small conducting tube into a convenient receiver filled with water and standing in the *pneumatic trough*. If the gas so collected is tested by means of a candle, having only a spark on its wick, the oxygen is readily recognized by the fact that the taper promptly bursts into a full and brilliant flame. This method is of historical interest chiefly, though it may well attract some attention from the simplicity of the chemical change involved. Thus this change is represented by the following equation:

$$2Hg O \text{ } heated \quad = \quad O_2 \quad + \quad 2Hg$$

Two molecules of Mercuric oxide, 432 parts by weight.	One molecule of Oxygen, 32 parts by weight.	Two atoms of Mercury, 400 parts by weight
432		432

A word about the pneumatic trough is not out of place here, because this useful contrivance was the invention of Priestley. The name may be appropriately applied to almost any vessel of water in which may stand the open mouth of a bell-glass suitable for containing gas. The water serves at once to seal the mouth of the jar, and also to afford a material through which the exit tube of an appliance may be dipped, and through which also the gas from the tube may freely and conveniently flow into the bell-glass. Before Priestley's time gases had been collected in bladders or varnished bags, but the new contrivance furnished a much superior means of detecting small quantities of gas and working with them.

Second Method of Preparing Oxygen.

The second method, and that oftenest pursued, employs a salt not known in Priestley's time. This salt is called potassic chlorate and is represented by the formula $KClO_3$.

This substance, when heated, evolves a large amount of oxygen, but it does so with almost explosive violence.

The chemical change is represented by the following equation:

$$2KClO_3 \text{ heated} = 2KCl + 3O_2$$

Two molecules of Potassic chlorate, 245 parts by weight.	Two molecules of Potassic chloride, 149 parts by weight.	Three molecules of Oxygen, 96 parts by weight.
245		245

On the other hand, if the potassic chlorate is mixed with about one-third of its weight of the earthy mineral known as black oxide of manganese, (but called by the chemist, manganese dioxide,) the mixture when heated evolves oxygen more slowly and continuously than the chlorate alone—and it does it at a lower temperature. Strangely enough however, the manganese dioxide appears to take either no *chemical* part in the operation or else only a very obscure one. Indeed, some other oxides will serve the same purpose, while they likewise appear to undergo no chemical change.

In this method, as in the other, the oxygen gas produced may be collected in a bell-glass over the pneumatic trough, and afterwards its nature may be demonstrated as before by means of the taper having a spark upon it.

The Properties of Oxygen.

It has been the custom of chemists to say of oxygen that it is *a permanent gas*. The force of this expression is found in the fact that until recently all attempts to liquefy it were futile. But recent experiments, with apparatus capable of subjecting it at once to more intense cold and to greater pressure than were ever before employed, seem to demonstrate that it will turn to a liquid when these conditions are carried to a sufficient extreme.

That oxygen is colorless and odorless appears plain from the properties of the atmospheric air throughout which this gas is thoroughly diffused and intimately intermingled, although it constitutes but one-fifth of it.

Chemical Properties of Oxygen.

Of the chemical powers of oxygen the most striking and important seems to be its marked tendency to combine with other elementary substances. In many cases this combination does not commence except when the substances are heated. Thus the noble buildings of a city are every day and every night continuously and harmlessly bathed within and without by that same oxygen, that, in time of conflagration, is ready chemically to combine with their elements and as a result to reduce the metropolis to ashes. But such combination, once inaugurated, often itself affords sufficient heat not only to make the process continue, but also to generate that flame or fire which is the token of what is ordinarily

FIG. 34.—The rays of sunlight concentrated, by a lens, upon a diamond placed in oxygen gas, with a view of proving the combustibility of the gem.

called combustion. In this view, oxygen is often spoken of as a supporter of combustion. That this property, known to be associated with the atmospheric air, does in fact reside in the oxygen of it, is to some extent proved by the more rapid and brilliant combustion of the candle in pure oxygen.

Another interesting experiment is performed when a piece of charcoal, which may be supported on a wire, is burned a little so as to acquire a spark, and then is dipped in oxygen gas. The single coal would soon cease to burn in atmospheric air, but it burns readily and brilliantly in pure oxygen.

Even the diamond, the most compact and imperishable form of carbon known, may burn in pure oxygen gas just as the most humble piece of coal does, and the relationship of the gem to the commonplace fuel is proved by this experiment.

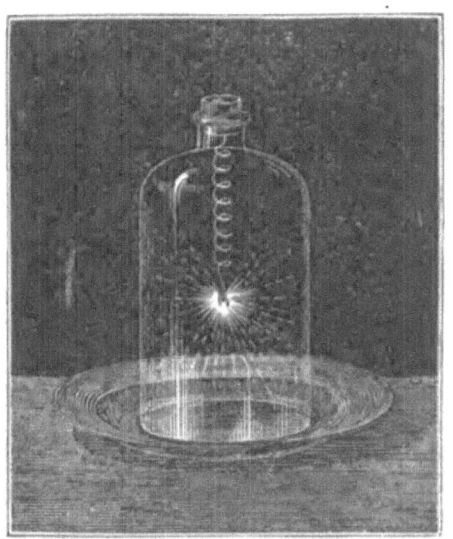

Fig. 35.—The burning of a spiral of iron wire in a jar of oxygen gas.

Still another experiment in the same direction may be conducted with sulphur. For this purpose a fragment of sulphur set on fire may be dipped in a jar of pure oxygen. The sulphur burns with vastly increased rapidity and with a violet flame much more brilliant than that of sulphur burning in air.

Again, some substances not ordinarily considered combustible will burn in oxygen gas. Thus a bundle of iron wire, to which a little lighted chip is attached, itself takes fire and burns brilliantly when dipped into oxygen gas.

The Products of Combustions in Oxygen.

As a necessary result of the combustion of substances in oxygen there are produced a multitude of compounds called oxides.

This is true of the candle, which consists mainly of carbon and hydrogen. When the candle burns, these two substances change into oxides. The carbon produces carbon dioxide, whose formula is CO_2, and which is familiarly known as carbonic acid gas. This oxide, it is true, is not easily recognized by the ordinary observer because it is an invisible gas, but the chemist can prove that it is in fact the product of this combustion. At the same time the hydrogen produces an oxide whose formula is H_2O and which will be recognized as the chemical expression for water. And so water is in fact produced, though in the form of vapor, by the burning candle.

Charcoal is composed almost entirely of what the chemist calls carbon, and when it burns it produces the oxide called carbon dioxide (CO_2). This is the same invisible gas that has already been declared to be produced when the carbon of the candle is burned, and in this case as in the other it is easy for the chemist to prove its presence.

In case of carbon, the chemical change is represented by the following equation :

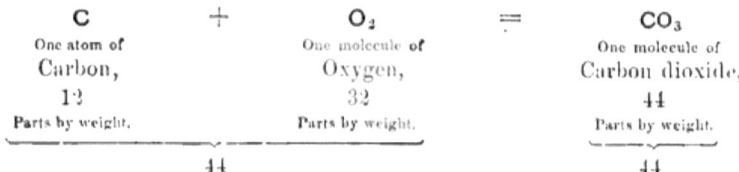

C	+	O_2	=	CO_2
One atom of		One molecule of		One molecule of
Carbon,		Oxygen,		Carbon dioxide,
12		32		44
Parts by weight.		Parts by weight.		Parts by weight.
44				44

And likewise when iron is burned, there is formed an oxide whose composition is expressed by the formula, Fe_3O_4; (to this substance the chemical name ferroso-ferric oxide is applied).

So when sulphur is burned, sulphur dioxide is formed (SO_2).

In this case the chemical change is represented by the following equation:

S	+	O_2	=	SO_2
One atom of		One molecule of		One molecule of
Sulphur,		Oxygen,		Sulphur dioxide,
32		32		64
Parts by weight.		Parts by weight.		Parts by weight.

64   64

Compound of Oxygen with Hydrogen.

It has already been shown that the hydrogen escaping from a suitable tube may be lighted in the air. If the burning jet is introduced into

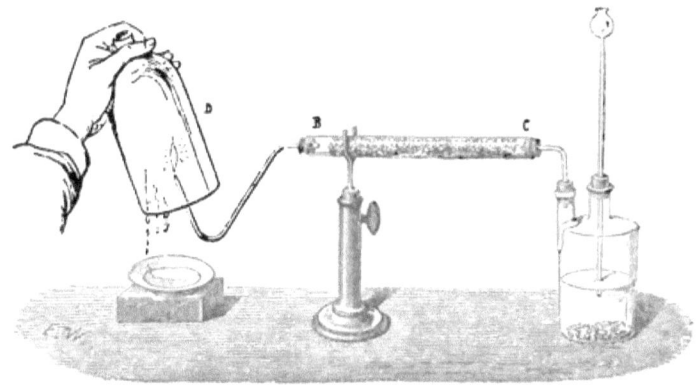

FIG. 36.—Hydrogen gas, generated by use of zinc and sulphuric acid, is then passed through a drying tube containing calcic chloride ($CaCl_2$). By the act of combustion the union of the dried gas with oxygen of the air produces drops of water.

oxygen gas the same combustion proceeds, only with greater energy. In either case there is produced a compound of hydrogen and oxygen. This compound is represented by the formula H_2O, a formula representing no other than the familiar substance water. At the moment of combustion of hydrogen very great heat is generated. In fact, a pound of hydrogen, upon burning in pure oxygen, yields about four times as much heat as a pound of pure carbon does in burning under the same favorable conditions. Indeed, the pound of hydrogen, when in combustion, yields more heat than a pound of any other substance known. On account of this heat the water resulting from the burning hydrogen at first floats off in the air in the form of vapor; but if the hydrogen flame is brought in contact with some cooling surface, the water formed condenses in drops upon it and thus it may be readily recognized as in its ordinary form.

A great multitude of experiments show that the composition of water is as follows:

Water is made up by the Union of,		
Parts by weight,	Parts by bulk,	Atoms.
Hydrogen, 2	2	2
Oxygen, 16	1	1

The composition of water, as displayed in the foregoing table, has been demonstrated by analysis, this word meaning "the process of taking apart." Thus by chemical influences a portion of water may be subdivided into its constituents and their amounts determined. On the other hand the composition of water has also been made out by synthesis, this word meaning "the process of putting things together." In this latter case, by putting together what are believed to be the proper proportional amounts of hydrogen and oxygen to form water, and then upon using some suitable means for bringing these things into a state of true chemical combination, it has been found that they do combine in fact to form water and in the proportions already given in the table.

The Compound Blowpipe.

The fact of the enormous heat developed when hydrogen burns, was known long ago, and it gave rise to the invention of a contrivance for utilizing it. This has taken the form of the apparatus called the compound blowpipe, also the oxy-hydrogen blowpipe.

This blowpipe, as usually constructed, has a single jet or tip— to which there is con, veyed by separate tubes, on the one hand oxygen, on the other hand hydrogen. The gases, when lighted,

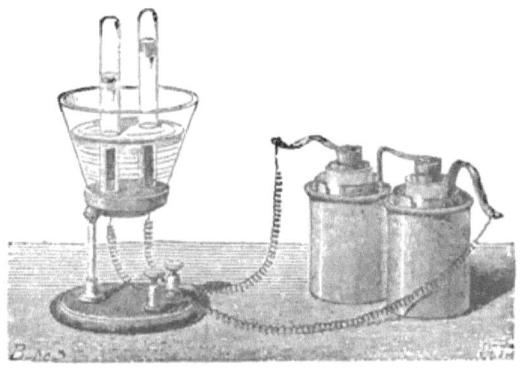

FIG. 37.—Apparatus for analysis of water by use of the galvanic battery.

give rise to a flame of but little luminous power but of intense heating power. Many difficultly fusible metals, such as iron for instance, melt like

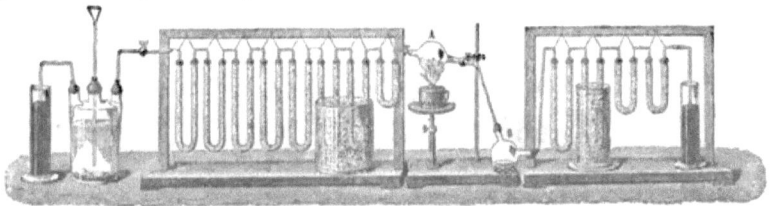

FIG. 38.—Apparatus devised by Dumas for determining the composition of water.

wax before it, while others, like lead and zinc, boil and vaporize beneath its fervent breath. It must not be looked upon however as a mere chemical toy; it has some uses in the arts. Of these one of the most prominent is its application to the melting and refining of the ores and alloys of platinum, substances which no ordinary furnace can liquefy.

For purposes of this sort, a special furnace or crucible must be provided, and it must be constructed of some substance that is itself practically infusible. Such a material is found in quicklime (calcic oxide, CaO), for this substance does not melt under the influence of any known contrivance for producing heat. Moreover it does not conduct heat rapidly, and thus any heat applied to the metal within, is not subject to serious loss by being conducted away through the walls of the vessel. For melting platinum then a crucible constructed of quicklime, and having a cover of the same material, is employed. A stream of burning gases from a compound blowpipe is forced through an aperture in the crucible cover in such a way as to fall on the metal to be melted.

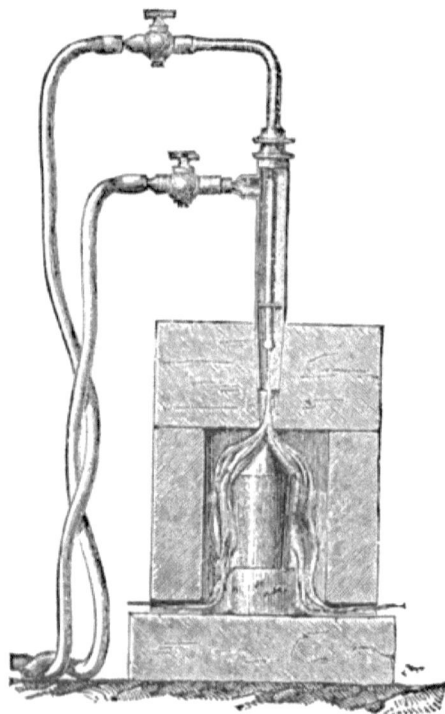

FIG. 39.—Flame of the oxy-hydrogen blowpipe directed upon a crucible in a furnace of lime.

The Calcium Light.

Another interesting application of this blowpipe is found in the lime light, an appliance also known as the calcium light and sometimes as the Drummond light. In this apparatus, whatever may be its particular form, the stream of burning gases is directed upon a small block or cylinder of lime. Of course the block becomes highly heated,—in fact it assumes a white heat, without melting; and while at this temperature it gives out a dazzling light. This light has been utilized by architects and engineers for carrying on important constructions during the darkness of night. It is also often used in some of the finer forms of the magic lantern, as for example in the various stereopticons used in illustrated lectures. So numerous are the uses of the calcium light in large cities that it has become a regular industry there to furnish the oxygen and hydrogen gases in separate iron cylinders or cans, into which they are pumped under great pressure.* (It is true that illuminating gas is sometimes substituted for hydrogen with decided economy in cost, and yet without serious loss of illuminating power.) When the cylinders are in use the stop-cocks are slightly opened, and the gases are under sufficient pressure to flow to the top of the blowpipe as freely as can be desired.

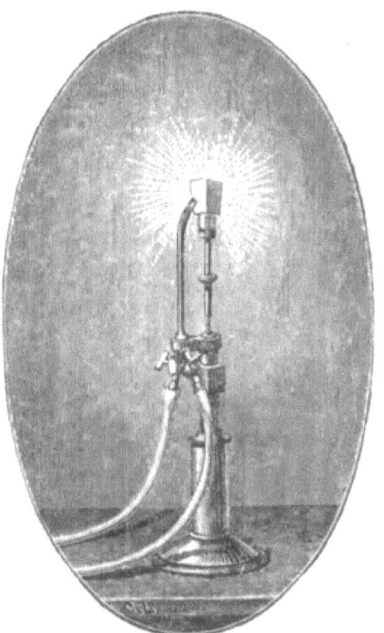

FIG. 40.—Drummond light, or calcium light. The flame of the oxy-hydrogen blowpipe directed against a block of lime renders the latter intensely luminous.

---

*DANGEROUS EXPLOSIBILITY OF MIXTURES OF OXYGEN AND HYDROGEN.—At this point a warning should not be omitted, for mixtures of oxygen and hydrogen gas, whether produced purposely or by accident, are capable of very dangerous explosions. Even a soap bubble, inflated with the mixed gases, and then lighted with a torch, explodes with tremendous violence and a loud report. This result is all the more wonderful when the extreme thinness and weakness of the filmy confining envelope is considered. Such explosions are in entire harmony with the various statements already made. For when the two gases combine, the intense heat then generated gives rise to a momentary but enormous expansion of the vapor of water produced by the combustion. The greatly expanded vapor immediately strikes the air a sharp and violent blow. In another instant however

Oxygen as Related to Combustions in General.

But oxygen is prominent in many other combustions besides that of hydrogen. Of course the best known and most common are those in which the ordinary forms of fuel are the things burned. Here generally the principal constituent of the combustible material is carbon.

Oxygen as Related to Animal Respiration.

Oxygen performs also one of its most important offices in connection with the process of animal respiration. In the fulfillment of this mission no element is known that can in any way act as a substitute. The gas, which is to serve as the breath of life for the humblest as well as the most exalted individuals of the animal creation, must possess a combination of qualities truly marvellous when residing in a single substance. Even a brief description of the ways in which it discharges this delicate and manifold duty ought to substantiate the general proposition.

Oxygen is qualified to sustain respiration by virtue of the exceeding abundance of the atmospheric air, an abundance such that it extends above our heads a distance of forty thousand times the height of a man. Nor are the denizens of the sea forgotton, for oxygen possesses such capacity for dissolving in water that there exists, absorbed in the liquid of the rivers and oceans, enough of this vital gas to furnish breath for all the finny tribes.

Again, the oxygen, so violent in its combinations, is yet bland enough to pass through all the delicate passages leading into the lungs without exciting the throat to the slightest cough ; to filter through the fine membranes of the lungs without doing an injury ; to saturate the blood, and to flow to every tissue and cell of the body, and not only do no harm but everywhere accomplish a reviving work. It performs throughout the animal frame a well regulated but no inconsiderable combustion. Indeed the

---

the vapor suddenly cools and condenses to an exceedingly minute drop of liquid water. Immediately upon this effect, the air that was previously forced outward immediately falls into the vacancy left, and now a second blow results. It is these two violent shocks the one following the other in almost instantaneous succession, that produce the report ; and to the same causes must be referred the terribly destructive results of the accidental explosion of considerable quantities of the mixed gases. It is plain therefore that all contrivances, destined to employ these gases in close proximity, must be handled with great caution when ready for use.

body of a living creature may be properly looked upon as a kind of furnace, taking in air whose oxygen shall sustain the combustion of worn-out parts. Nay more, these as they burn do in their very death make as a final contribution the gift of that warmth and glow which maintains the animal temperature at the vital point.

While carrying out the important functions just referred to, oxygen produces several gaseous substances, each of which, as if under the constant direction of an ever watchful barometer, maintains its proper bulk and pressure, so as to do no injury to the most delicate capillary of a vein or to the tender walls of the smallest chambered cell of the lungs. With each breath exhaled from the system, the blood, and thence the lungs, discharge the gaseous products of the combustion already described; plainly they do it somewhat in the same manner as a chimney does in its proper action, only the lungs do their work in a far more perfect way.

The parallelism is not strained here, for the burning of the animal tissue in the body gives rise principally to the production of the gas called carbon dioxide and the vapor of water, just as when a faggot burns in the chimney-place, the carbon and the hydrogen of the wood oxidize into the selfsame products, both of which are wafted up the flue and out into the great ocean of atmosphere beyond.

READING REFERENCES.

Gases, Liquefaction of
  Cailletet, M.—Annales de Chimie et de Physique. 5 Sér. xv, 132.
  ——Chem. News. xxxvii, 11.
  Coleman, J. J.—Chem. News. xxxix, 87.
  Pictet, R.—Annales de Chimie et de Physique. 5 Sér. xiii, 145.
  ——Chem. News. xxxvii, 1, 23, 83.
  Roscoe and Schorlemmer.—Chemistry. New York. 1878. ii, pt. II, 516.
  Schutzenberger, P.—Traité de Chimie Générale. i, 25.

Priestley, Joseph
  Brougham, H.—Lives of Men of Letters, etc. p. 402.
  ——Amer. Chemist. iv. 362-441; v, 11-35, 43, 210.

CHAPTER XV.

WATER.

AS the most prominent compound of oxygen, water may properly receive the reader's attention at this time.

He who stands upon a high cliff and looks out upon the ocean, experiences as one of his strongest impressions that of the boundlessness of the expanse. And it is true that the area of terrestrial waters is very wide, for in the aggregate their waves cover more than three-fourths of the earth's surface. But while their superficial extent is so great, their depths are relatively but small. When compared to the diameter of the earth the deepest ocean seems shallow indeed. If the waters of the oceans were dried up or otherwise wiped away, the roughness of the dry globe would be less relatively than the roughness of an orange. In fact the total amount of water actually existing upon the earth's surface is less—relatively to the entire mass of the globe—than the amount that would remain on an orange after dipping it into a basin of water and then withdrawing it. Notwithstanding these facts, the amount of water is so vast in proportion to the littleness of human beings, and it has taken so prominent a part in the phenomena observable by man, and it has been such a powerful agent in the geological eras of the past, that it is not surprising that its properties and history have excited the interest of students and thinkers of all times. In the light of modern chemical knowledge too, its various offices create an admiration that is heightened, the more they are considered.

The chemical history, the characteristics, the properties, and the uses of water, all these are important chemical topics; when one considers further, the varied forms and uses in which this familiar substance is employed in nature and in the arts, a subject is suggested that might well furnish material for a volume. Plainly then only a few of its more striking adaptations can be discussed here.

Importance of Water to Living Beings.

To living animals and plants water appears to be absolutely indispensable. The reason for this is found not only in the fact that water forms a necessary constituent part of most living beings, but also because it serves as a sort of vehicle by virtue of whose properties the vital processes are conducted and through which the vital currents flow. It is easy to understand that if the atmospheric air, which lies wrapped about our globe like a thin veil, were suddenly wafted away, animal life would be instantly extinguished. Now, water is not less essential than air. Banish water from the earth, and the life of all animal and vegetable beings would instantly take its flight. For the blood, that living tide which courses through the natural gates and alleys of the body, contains water to the extent of nearly seven-eighths of its weight. Again pure, unadulterated milk, rich as it is in solid food materials dissolved or suspended within it, contains not far short of 90 per cent of water. And further, an examination of vegetable products reveals in them a preponderance of water such as would not at first be suspected. Thus the following brief table represents facts so surprising that it is at first difficult to accept them:

Fig. 41.—Egyptian Water-carrier.

Apples   contain about   80   per cent of water.
Turnips   "   "   90   "   "   "   "
Cucumbers   "   "   97   "   "   "   "

Finally as an extreme example among the kingdoms of life it may be mentioned that some forms of jelly-fish, as taken from their appropriate home in the ocean, have been found to contain not less than 99 9-10 per cent. of water.*

The extraordinary and incredible proportion of water in living beings is associated with the numerous, varied, and even apparently contradictory offices to be performed by it, and the fitness of water to fulfil these requirements is referable further to the curious and interesting properties with which it is endowed. But it is so familiar to every one and so bland in its action in its relation to most well-known substances, that the ordinary observer fails to recognize these properties and their marvellous adaptations.

One of the properties most appropriate for presentation in this connection is the power water possesses of dissolving gases. It is capable of storing up within itself, concealed from human view, almost every gas with which it comes in contact. It displays this power upon the atmospheric air, not only in its better known relations to man and the higher animals, but also as respects the humbler population of the globe. It will be seen by-and-by that the air consists in the main of a mixture of two gases very different in their properties. One is oxygen, the sustainer of animal respiration; the other, nitrogen, the inactive substance existing in the air as a mere diluent of the active oxygen. Now water possesses a very curious relation to these gases; it naturally dissolves a larger proportion of oxygen than of nitrogen. By reason of this property it acts upon the atmospheric air with a selective effect highly suggestive of intelligent plan. For the gas it selects to dissolve in larger proportional quantity is oxygen,—the one absolutely needed to take the principal part in supporting the respiration of the countless millions of fishes that make their natural homes in all great bodies of water.

Terrestrial Circulation of Water.

Water is the chief liquid of the great globe itself. And it carries on here a continued and beneficent circulation which may be properly likened

*COOKE, JOSIAH P.: *Religion and Chemistry.* New York, 1864. p. 148.

FIG. 43.—Arctic explorers employing dynamite to open a channel through the ice.

to that of the living animal and plant, except that it proceeds on the cosmical scale. This circulation may be described as starting in the depths of the ocean, where permanent currents are constantly flowing in certain directions. These contribute to make the seas the highways of navies even more completely than they would be if the waters were always at rest. A yet more striking circulatory movement is that initiated by the volumes of moisture which rise by constant evaporation from the temperate as well as the tropical seas. This water, ascending into the higher atmosphere, is carried by currents of the air hither and thither and over the land, where by mountain ranges or other natural means adequate to this purpose, it becomes precipitated into a solid or liquid form. In this condensed form it is recognized as beneficent when it is in cloud masses which delight mankind with the purity of their fleecy whiteness, or the beauty of their gorgeous coloring as well as when it is in the form of showers which refresh the thirsty earth, or as the snow which protects it. The rain and snow supply the numberless rivulets that contribute to make up rivers, and these flow joyfully to the ocean and, mingling in its waters, return to the source from which they came. Thus has been pictured in brief an outline of the circulation previously suggested.

Water in the Solid Form.

Again, certain properties of water in the solid form are worthy of presentation. Perhaps it is not inconsistent with the truth to say that they are even more plainly beneficial. Thus in the form of snow, water appears at first sight to be an emblem of cold. But when it falls upon the earth it becomes a mantle or coverlet, which protects the soil from the chilling effects of the wintry season and from that rapid loss of heat by radiation off into space which the fields would suffer without this protective coating. And so ice, as it forms on the surface of lakes and ponds, manifests several remarkable properties. Of these only two will be discussed here. They are both due to its power of expanding at the moment of solidification. Most persons make acquaintance with this characteristic of water by the inconvenient bursting of pitchers and pipes, recognized as a disagreeable attendant upon the winter's cold. When looked upon with more fully instructed eyes, however, it is discovered to be one feature of a remarkable system which results in great benefit to the inhabitants of the earth. For it is plain that as water in freezing expands, it thereby becomes relatively lighter. On this account ice floats in water, whereas solid substances generally sink in liquid matters of their own kind. Now the ice formed upon lakes in the winter, stays at the top and thus protects the water below from the chill of the colder air; so it prevents the lakes from becoming uninhabitable to the fish. The same property prevents a lake from becoming a mass of solid from the bottom upwards, as would be the case if the ice upon freezing went to the bottom. The summer's sun would hardly be capable of thawing the solid masses so formed. This same curious fact of the expansion of ice at the moment of its formation contributes to the fertility of the soil. Thus the water that penetrates the crevices of rocks, expands upon freezing, chipping off those rocks, in fact pulverizing them little by little, and so conveying fresh and valuable materials to the earth's soils.

Fig. 42.—Water in the form of cumulus clouds.

Water as Affecting Climate.

Further, the relations of water to heat are very interesting. "The general aqueous circulation of the earth is a great steam-heating apparatus, with its boiler in the tropics and its condensers all over the globe. The sun's rays make the steam. And wherever dew, rain or snow fall, there heat, which came originally from the sun, and which has been brought from the tropics concealed in the folds of the vapor, is set free to warm the less favored regions of the earth. This apparatus in nature, although so much simpler and working without pipes, iron boiler or radiators, is exactly the same in principle as the steam heater which may be seen at work in almost every large factory."* In other words, when water is changed into vapor in the tropics, heat is not only requisite to the operation, but a definite quantity of heat is actually stored up within the vapor so produced. On the other hand, whenever in some cooler parts of the globe this same portion of vapor condenses into the form of liquid, that heat that was stored within it at the tropics is, immediately evolved and contributes something to the warmth of the region where condensation takes place. Nay more, if the water, instead of falling as rain, falls as snow a still larger amount of heat is by this means given out into the atmosphere. This last statement is insensibly substantiated by the expression often heard in winter, "the weather is too cold for snow." This common expression, translated into scientific language, means "the air does not possess that amount of warmth that it would manifest if snow were now condensing in the upper air and were ready to fall."

It is not only with respect to those changes taking place when the vapor of water changes to the liquid or the solid form that its heat relations are beneficial to mankind. No lake can change one degree in temperature— that is, grow warmer or cooler—without at the same time exercising a contrarywise influence upon the air about it, and thus a regulating one. In explanation of this declaration the following statements may be made: When, in the intensely hot days of summer, a lake or any mass of water become influenced by the high temperature, of course its waters become warmer. But it is a curious fact that it takes more heat to raise the temperature of water one degree than it does to raise the temperature of the adjoining land one degree—or in fact to raise any other substance known, one degree. Thus it appears that a given amount of heat applied in a summer day to a lake will be absorbed within the waters of

---

*COOKE, JOSIAH P.: *Religion and Chemistry.* New York, 1884. p. 135.

that lake without raising the *temperature* of those waters to the extent that might be expected. So then in hot weather the lake becomes an equalizer of temperature with a tendency in the opposite direction, that is to cool the air about it. Now in cold weather it becomes equally beneficial, only, as might be expected, in the opposite direction. Thus the store of heat retained by the liquid water is given out as the lake cools. For just as the water in order to rise one degree in temperature requires, and indeed absorbs, more heat than any other substance known, so naturally the same water, in cooling one degree in temperature, freely gives out the amount of heat it had previously stored within itself, which, as has been said is greater than that stored up by any other substance known.

Water as a Working Contrivance.

When the moisture of the tropical oceans is taken up into the air by evaporation, the sun has thereby done a truly stupendous amount of *work*. For has it not lifted up high into the atmosphere an enormous weight of this liquid material? Now as the vapor is wafted over the land preparatory to falling as rain, it has acquired a position in which it may do a great amount of work for human uses; for every rain-drop, falling from its lofty position in the air, acquires thereby a momentum which represents a quantity of force, minute in each individual case but truly vast in the aggregate. Of this sum total but a small portion is employed for man's industrial uses; only a minute fractional part is harnessed to the wheels that grind his food or weave his clothing or transform the trees of the forest into his habitations; yet the amount he does so employ—compelling it to do his work for him—represents an enormous total quantity. All this work done, as well as all that might be done, by the vast quantities of water allowed to escape and violently run to waste, is referable back again to the sun of the tropics, which has been enabled, by reason of the wonderful properties of water, to store up all this power within it.

In view of what has been said, the sun and the water of the tropics may be compared not inappropriately to the chief artificial contrivances used in modern times for generating and applying mechanical power—boiler and engine. As an ordinary steam-boiler imparts to water the expansive and working power of steam, and again the ordinary steam-engine utilizes this steam power so that it may be directly applied to the labor of man's workshops, so the sun of the tropics lifts the water of the ocean high up into the air, and thus may be likened to the boiler; while the rapidly running brooks may fitly represent an engine in motion, ready to actuate any machine to which by proper appliances it may be attached.

CHAPTER XVI.

SULPHUR.

SULPHUR, in its aggregate in the earth is by no means an abundant element. Thus its quantity is far inferior to that of oxygen, as is strikingly illustrated by the diagram already presented. (See page 16.) Yet sulphur was recognized by human beings thousands of years before oxygen, which it has already been stated was discovered in 1774. The comparative lateness of the discovery of this latter element, now known to be that one which predominates largely over all others in the earth, is due partly to the fact that free oxygen almost invariably exists in gaseous form and that the idea or notion of gas is one of recent growth. The fact that sulphur was recognized so much earlier is due to many circumstances. *First*: It is found in the earth in the solid condition—a form at once tangible and easy of recognition. *Second*: Its yellow color helps to render it noticeable. *Third*: It exists in the earth in countries which have long been the abode of civilized beings. Thus it was early recognized in Italy. *Fourth*: It occurs in deposits of such a character that it can be readily obtained in a comparatively pure form from them. *Fifth*: It possesses certain remarkable properties some of which would be easily detected even by savage peoples, while others have for centuries excited great interest in the minds of students of alchemy and chemistry. One of these properties is the ease with which it assumes a liquid form—that is, melts—when slightly heated. Another is the readiness with which it takes fire and burns in the air. A third, closely connected with the foregoing, is the striking blue flame produced when it burns.

Still another, and not less noticeable, is the choking and disagreeable odor attendant upon this combustion.

Finally may be mentioned a circumstance which for a long time contributed to make it peculiarly interesting to the alchemist, if not to ordinary men: this is the fact that when sulphur is in the pure form it may be burned away without leaving any ashes. In this respect it differs from most other combustible materials. And this property created the impression that sulphur is a sort of principle of fire, and that it somehow exists in all combustible bodies. Indeed it is only for about a hundred years that sulphur has been classified as a distinct elementary form of matter. It is not intended to indicate here that the strong interest of the alchemists in sulphur was mainly referable to the circumstances of its combustibility. Its power of combination with the metals was well known to them, and was recognized as a subject of practical importance and one worthy of careful study and thought.

Natural Sources of Sulphur.

The principal supply of sulphur for commerce is obtained from the volcanic districts of the island of Sicily. Here in fact there are more than two hundred distinct establishments for production of the substance, and they are capable of yielding about two hundred million pounds of it per year.

The fact that sulphur is easily and widely recognized in the earth has already been dwelt upon. But it occurs in nature in a great variety of forms. The first and most striking form is that of free and uncombined sulphur. In this condition it occurs either as masses or as fine powder. Sometimes these materials possess the well-known and easily recognized yellow color of sulphur; sometimes however the color is white or otherwise disguised by reason of some peculiarity of the sulphur itself or else because of the admixture of foreign substances. Deposits of sulphur occur in the most considerable quantities in the neighborhood of either active or extinct volcanoes. Thus sulphur earth occurs near Vesuvius and Ætna, also in the vicinity of the volcanoes of Iceland, in Central America and in the Sandwich Islands. In the region of some extinct volcanoes the soil is impregnated with sulphur to the depth of twenty or thirty feet and such soil is therefore a convenient source of the element.

Purification of Natural Sulphur Ores.

In obtaining sulphur from the earth for commercial purposes, two simple processes are resorted to. By the first method masses of the sulphur earth are heaped up into a pile, in connection with a small amount of fuel and over a shallow depression in the earth. Upon setting the mass on fire,

Fig. 44. *Calcarone* or heap of burning mineral from which sulphur is obtained.

considerable quantities of sulphur escape combustion, and so melt and run down to the ground below the heap. When the fire is extinguished, the sulphur that collected beneath may be secured in a form now only slightly impure.

The second method of purification of the earth is still conducted in Sicily in the following crude manner, though this is quite an improvement upon that just described. A slightly inclined plane of masonry is built upon the ground. Around the edges of this plane a low wall is erected. At the lower side of the plane the wall is perforated. Upon the surface of the plane large masses of sulphur earth are carefully piled up so as to form

Fig. 45.—Sketch illustrating the process of refining sulphur.

a well-built heap. When it reaches the proper height its outside is covered all over, first, with small fragments of the same kind of earth, and then with its fine dust. Some sulphur at the lower portion of the heap is then set on fire at several points. The heat from the sulphur that burns melts other portions of it, which then trickle down the spaces between the masses of rock. This melted material, finding the bottom of the pile, runs freely to the lowest portion of the platform, then through the perforations and out into wooden boxes placed to receive it. The heap burns for two or three weeks, at the end of which time the operation is finished. When the mass is cool it is torn down, and a similar pile is erected from fresh portions of the sulphur earth. The objectionable features of this process are at least four. First, the consumption of sulphur as fuel is a wasteful one. But in reply it may be said that no cheaper fuel is accessible where this manufacture is carried on. Again, the great volumes of sulphur dioxide given out by the burning *calcaroni*—as the heaps are called—are injurious to the health of the workmen. Further, these same products exercise a very destructive effect upon all vegetation in their vicinity. In fact on this account the Italian government has provided by law that this work shall not be carried on at all between July 1st and December 31st. Finally the method is not as successful with the richer ores, for they break down into powder which it is difficult to utilize in the calcaroni.

A new and greatly improved method, and one which overcomes all the objections above cited has recently been introduced. In this, the ore is placed in perforated metal baskets and then immersed in tanks containing hot solutions of calcic chloride. Under these conditions the sulphur melts out from its ore and falls to the bottom of the tanks, whence it is drawn out by stopcocks in a comparatively pure form.

Sulphur is generally subjected to a still further purification. This is conducted somewhat as follows. The crude sulphur, being melted in a suitable retort and over a coal fire, changes into vapor and passes into an apartment constructed of stone or brick, and prepared for the purpose. In this apartment, the sulphur at first condenses on the walls as minute yellow crystals or powder called flowers of sulphur. When the first charge of sulphur in the retort has been completely vaporized a new supply is allowed to run in, this time in the liquid form from a small heater placed above the retort. The waste heat from the furnace melts the sulphur in the heater, from which it flows into the retort (by means of the tube shown in the diagram). When a sufficient amount of flowers of sulphur has collected in the chamber, the fire is extinguished. The purer product is then removed. Afterward the whole operation is repeated.

The refining may be conducted so that the temperature of the condensing apartment may rise considerably; in this case the vapor in it changes to the liquid form. This liquid may be drawn off at the base of the chamber into a small receiver, from which it is ladled into moulds, which give it the form of cylinders known in trade as roll brimstone.

Natural Compounds of Sulphur.

Sulphur is also found in the earth in the form of certain chemical compounds. Some of these are very widely distributed. They may be divided into two classes. The first class—whose representatives are by far the more abundant—includes the metallic sulphides, that is, compounds formed by the direct union of sulphur with some metallic substance. As examples of compounds of this class we mention:

Sulphide of iron (commonly called iron pyrites and having the formula FeS_2).

Sulphide of lead (commonly called galena, and having the formula PbS).

Sulphide of zinc (commonly called blende, and having the formula ZnS).

Sulphide of mercury (commonly called cinnabar and having the formula HgS).

Many other examples of similar import might be given, for it is a well-known fact that most of the heavy metals occur in the earth in combination with sulphur.

The other class of compounds also containing sulphur combined with the metals has usually oxygen in addition. Two examples of this class may be given here—calcic sulphate (commonly called *anhydrite*, and having the formula $CaSO_4$); also baric sulphate (commonly called *heavy spar*, and having the formula $BaSO_4$).

Sulphur is very widely distributed in animal and vegetable matters. In these it exists, not as an uncombined element, but in union with others.

Indeed such compounds have many other elements besides the sulphur, and they are characterized by decided complexity of structure. But sulphur is oftener a component of animal matters than of vegetable. The presence of sulphur in an egg is proved by an experiment of every-day occurrence. That is to say, the silver spoon with which the egg is eaten becomes blackened. This blackening is due to the production of a new compound formed by a true union of sulphur from the egg, with a part of the metal of the spoon. In fact the black material is sulphide of silver, and it may be represented by the formula Ag_2S. A French chemist has estimated that in the body of a human being of ordinary size there exists, in the aggregate, not far from one quarter of a pound of sulphur. To this, he adds the curious estimate that the entire human population of France may be represented as containing not far from nine millions of pounds of sulphur.

Chemical Properties of Sulphur.

The chemical properties of sulphur may be said to be its most important and interesting ones. That it has a wide range of chemical aptitudes is shown by the fact that it combines in simple forms of union with a majority of the elements known. Thus it has strong affinities for most of the metals. On the other hand it combines with various degrees of attractive force with nearly all the non-metals as well.

Evidently then, sulphur forms a very large number of chemical compounds. While the limits of this work are such as to make it impossible to describe many of them, there are three that may with propriety be briefly discussed in this place, and these are:

Sulphuretted hydrogen (H_2S),
Sulphur dioxide (SO_2),
Sulphur trioxide (SO_3).

Sulphuretted Hydrogen.

This substance is a colorless gas. It has an extremely offensive odor; in fact it is a prominent component of that numerous group of gaseous products of decomposition of animal matters that produce the disagreeable smell attendant upon the decay of the latter.

Again, it is found in the waters of certain natural sulphur springs, and it is a remedial agent of considerable value when properly applied externally or when taken into the stomach. When received into the lungs, however, it is decidedly poisonous.

A considerable number of simple experiments may be tried with it.

In these the gas used is generated by adding diluted sulphuric acid to artificial ferrous sulphide. The ferrous sulphide is usually manufactured by heating a mixture of roll brimstone and iron filings in a sand crucible. In producing the gas, the chemical change is represented by the following equation:

FeS	+	H_2SO_4	=	H_2S	+	$FeSO_4$
One molecule of Ferrous sulphide,		One molecule of Sulphuric acid,		One molecule of Sulphuretted hydrogen,		One molecule of Ferrous sulphate,
88		98		34		152
parts by weight.		parts by weight.		parts by weight.		parts by weight.

186   186

For the purpose of the experiments here mentioned, a flask or bottle may be used to prepare the gas and convey it into another bottle containing water. In the water, the sulphuretted hydrogen gas dissolves in such quantity that the solution so afforded may be conveniently employed for showing the properties of the gas itself.

The following interesting experiments may be performed by use of this solution:

1. Dissolve in water a small quantity of plumbic acetate, also called sugar of lead. Filter this solution if convenient. To the clear liquid, add some sulphuretted hydrogen water. A black precipitate of plumbic sulphide (PbS) should immediately appear.

2. Dissolve in chlorohydric acid a fragment of white-arsenic not bigger than a pin's head. To the solution, freely add sulphuretted hydrogen water. A beautiful lemon-yellow precipitate, consisting of arsenious sulphide (As_2S_3), should result.

3. Dissolve in chlorohydric acid a minute quantity of tartar-emetic. To the solution, freely add sulphuretted hydrogen water. A beautiful orange red and flaky precipitate of antimonious sulphide (Sb_2S_3) should appear.

4. Dissolve in water a minute fragment of cupric sulphate, com-

monly called sulphate of copper or blue vitriol. To the solution, add some of the sulphuretted hydrogen water. This should instantly give rise to a black precipitate of cupric sulphide (CuS).

5. Dissolve in water a small quantity of zinc sulphate. To the solution, freely add sulphuretted hydrogen water. There should appear in this case a white precipitate consisting of zinc sulphide (ZnS).

These few experiments show that sulphuretted hydrogen is a convenient substance for bringing sulphur into union with the metals, and, moreover, they sustain the statements already presented, that many metals show strong affinity for sulphur and marked tendencies to combine with it. For these reasons sulphuretted hydrogen is much used in chemical laboratories for distinguishing one metal from another.*

Sulphur Dioxide.

When sulphur burns in oxygen gas or in atmospheric air, it gives rise to a new gas of choking and offensive odor. This is the same substance as that produced in the first stages of the burning of a sulphur match. It is a substance of considerable importance in the arts, first, because it is always produced in one stage of the process used in the manufacture of sulphuric acid. Now sulphuric acid (commonly called oil of vitriol) is a commercial product of enormous consumption. (See page 152.) Again, sulphur dioxide is used, as such, to a considerable extent in the arts, the principal uses being in the bleaching of straw and woolen goods. Chlorine as a bleaching agent has already been discussed, but it is used mainly for the bleaching of cotton and linen goods; it has an unfavorable and injurious action upon straw and woolen goods.

The way in which these latter are bleached by the use of sulphur may be illustrated by a very simple experiment. Place a few fragments of roll brimstone in a small crucible. Heat the crucible carefully until the sulphur takes fire. Then cover the burning sulphur with a glass lamp-chimney, or any other suitable contrivance. In the top of the chimney hang a moistened carnation pink or other red flower. A few minutes exposure to the gas, results in a partial bleaching of the flower.

---

*See Appleton's Qualitative Analysis, published by Cowperthwait & Co., Philadelphia; p. 14.

On a commercial scale, the sulphur bleaching process is conducted in practically the same manner. For bleaching woolen goods there is provided a small wooden house having a brick floor, with a small pit in the centre. The goods are hung up in this house. The pit is filled with sulphur which, when all is ready, is set on fire by throwing a piece of red-hot iron upon it. Now the doors and windows of the house are closed. Of course the sulphur burns into sulphur dioxide. The operation is allowed to proceed without any further attention during one night. The gas distributes itself throughout the goods and bleaches them. The next morning the doors and windows are opened, and, when the fresh air has driven the sulphur dioxide from the chamber, the goods are found bleached. Everyone knows, however, that this bleaching has not the permanence that chlorine bleaching has. Thus white flannels very soon return to their original yellowish shade.

Sulphur dioxide is placed by the chemist in the class of acid *anhydrides*. This term is intended to carry the meaning that substances belonging to this class combine with water to form acids. In accordance with this form of expression, sulphur dioxide is also called *sulphurous anhydride*. Plainly this means that sulphur dioxide with water will form an acid. Such seems to be indeed the case, for water has the power of dissolving large quantities of sulphur dioxide, and when it does so the water acquires the characteristics of an acid. In fact it is then called sulphurous acid. The chemical change is represented by the following equation:

$$SO_2 \quad + \quad H_2O \quad = \quad H_2SO_3$$

| One molecule of Sulphur dioxide, 64 parts by weight. | One molecule of Water, 18 parts by weight. | One molecule of Sulphurous acid, 82 parts by weight. |

$$\underbrace{}_{82} \qquad \underbrace{}_{82}$$

One special characteristic which justifies the name sulphurous acid, is the fact that the solution so produced has the power of producing a series of salts as the other acids do. In this case, the salts have the general name sulphites.

CHAPTER XVII.

SULPHUR TRIOXIDE.

SULPHUR trioxide does not exist by itself in nature. Moreover it is but little known even as an artificial product. It is not an article of ordinary sale, though it is occasionally made by the chemist. Yet it is a constituent part of one of the most important compounds known to modern industry. That compound is sulphuric acid.

Sulphur trioxide is a white solid, but it cannot easily be kept so. This is because it has very strong affinity for moisture. It fact it readily absorbs that water vapor which is distributed through the atmosphere, even in dry weather and when the ordinary observer would suppose that the air contained no moisture at all. When it absorbs moisture it chemically combines with it, forming sulphuric acid.

The chemical change is represented by the following equation:

$$SO_3 + H_2O = H_2SO_4$$

| One molecule of Sulphur trioxide, 80 parts by weight. | One molecule of Water, 18 parts by weight. | One molecule of Sulphuric acid, 98 parts by weight. |

98 = 98

On account of this reaction, sulphur trioxide is often spoken of as sulphuric anhydride, the term anhydride being intended to suggest that the substance so named is derived from an acid by the removal of water from the latter. Thus sulphuric acid *minus* water produces sulphuric anhydride. And this harmonizes with what has before been declared, namely, that sulphuric anhydride—or sulphur trioxide—*plus* water produces sulphuric acid.

Sulphuric Acid.

This substance is known to commerce chiefly under the name of oil of vitriol. It is an oily liquid nearly twice as heavy as water. It has very powerful chemical action upon most substances with which it comes in contact. Moreover, its market price is very low, that is, between one and two cents a pound at wholesale. To these two facts last mentioned—that is, the marked chemical power and the low price is referable the enormous demand for the substance. To be sure, increase of demand and fall in price have a reciprocal action; for even a slight cheapening of a substance widens considerably the range of its possible uses and increases the amount consumed. Again, increase of demand and consumption, lead manufacturers to increase their production, a circumstance which is generally followed by lower price. The manufacture of sulphuric acid exemplifies these well-known principles of political economy. The manufacture of this substance has risen within the last hundred years from almost nothing to a present annual production of about nine hundred thousand tons in Great Britain alone. The price meanwhile has fallen to about one-thirtieth of what it was in the middle of the last century. At the present time the price of oil of vitriol seems to be steadily decreasing, while the amount produced is steadily increasing in England, France, Germany and the United States—indeed in all countries pervaded by active industrial enterprises. It will be generally admitted, as M. Dumas has said, that the amount of sulphuric acid consumed affords a very precise measure of the advancement in industrial arts of a given country or of a historical epoch.

Uses of Oil of Vitriol.

It would be difficult to enumerate the many industries that demand the use of sulphuric acid. It must likewise be admitted that there are but few manufacturing operations which do not directly or indirectly involve

Fig. 48.—Section of chambers for manufacture of sulphuric acid.

its employment. The industries that stand in the front rank as direct consumers of this acid are those that involve the following processes, namely: the bleaching of cotton goods; the removal of scale from iron in its various forms, such as castings, wire, etc.; the changing of corn starch into the variety of sugar commonly called glucose; the refining of bullion of gold and silver; the refining of petroleum oil; last, but not least, the manufacture of chemical fertilizers for agricultural use. Less directly, but still in enormous quantities, it is used in the manufacture of soda-ash, and bleaching powder already referred to as having reached an incredible consumption; in the manufacture of alum; in the manufacture of both of the great acids of commerce, chlorohydric acid and nitric acid, which must be said to come next to sulphuric acid in usefulness; and finally, in almost all the distinctly chemical industries.

Manufacture of Sulphuric Acid.

Notwithstanding the extremely low price of oil of vitriol and the immense quantity of it manufactured, its production implies a series of processes far more complicated than those invoved in the preparation of any other well-known acid. Moreover, although the various intricate details of its preparation are matters of thorough *experimental* knowledge to the producer, there are several steps which are not yet clearly comprehended even by the most eminent chemists of the age.

The process of manufacture, as at present conducted, is properly described as a continuous one. By this it is meant that the raw materials are steadily introduced at one end of the apparatus used, and the finished product is steadily drawn out at the other, the process meanwhile going on without interruption, night and day, for years. In order to a better comprehension of the process it is here described in four stages.

In the first stage, sulphur is burned in a current of air. The material employed is either partly refined Sicily sulphur, or what is largely used at the present day, some mineral compound of sulphur, like the iron and copper pyrites. In either case, sulphur dioxide (SO_2) is formed. This is the well-known choking gas given out by a burning sulphur match. As produced on a large scale the gas passes into a series of enormous leaden chambers. These are, in fact, rectangular rooms, often as large as one hundred and fifty feet long, twenty feet wide and fifteen feet high. Generally at least three chambers are in a series, connected by leaden pipes Sulphur dioxide gas flows in a steady stream into the series of chambers and toward the high chimney of the works, whose draft produces the advance of gases through the whole apparatus.

The second stage is the most complicated one. It is the oxidizing of the sulphur dioxide (SO_2) into sulphur trioxide (SO_3). This is indeed accomplished by means of the oxygen of the air But this oxygen is not

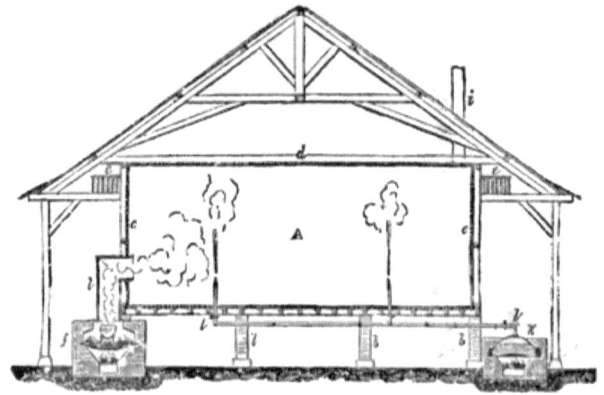

Fig. 46.—Section of building fitted for manufacture of Sulphuric acid; f, furnace where sulphur is burned and oxides of nitrogen are liberated; k, boiler from which steam is supplied to the leaden chamber, A.

capable of *directly* changing SO_2 into SO_3. Certain gaseous oxides of nitrogen are forced into the chamber at this stage; and these have the remarkable power on the one hand of taking oxygen to themselves from the air, and on the other of imparting this oxygen to the compound SO_2 in such a way as to change it into the the compound SO_3. Of course the air is impoverished by the operation, a fact which necessitates a fresh supply of it through the entire series of chambers.

The third stage is one whose principle has already been explained. At various parts of the chamber, jets of steam are blown in. While these aid mechanically in the progress of the gases through the entire series, their main purpose is to furnish water which shall combine with sulphuric anhydride to produce sulphuric acid.

Although this chemical change, represented by the following equation, has been given before, it may not be improper to repeat it here:

SO_3	+	H_2O	=	H_2SO_4
One molecule of Sulphur trioxide,		One molecule of Water,		One molecule of Sulphuric acid.
80		18		98
parts by weight.		parts by weight.		parts by weight.
98				98

The effect of the steam is to give rise to a steady rain of oil of vitriol in the chambers. Of course this liquid collects at the bottom. Thence it is drawn off, for treatment in a fourth stage. It is plain that up to this point the series of chemical reactions takes place in what we may characterize as a vast but irregular tube, open at both ends. This tube is enlarged here and there into great pockets which constitute the chambers. It is bent into a form appropriate to the conditions of the business. It is entered here and there by pipes for introducing the agents whose proper interaction gives rise to the product sought. It is also tapped for the purpose of drawing off the acid generated. This open tube has its final exit into the atmosphere through the tall chimney with which it is connected. It has its first connection with the atmosphere at the open throat, which swallows at once the vast volumes of sulphurous gas from the sulphur burned, and at the same time levies upon the air to contribute its oxygen to produce the substance which is the final purpose of the whole industry.

The fourth stage is the only one that may be properly said to be disconnected from the others. The continuous process already described cannot properly be made to produce acid of the strength demanded by commerce. In the fourth stage then, the acid from the chambers is boiled with a view of expelling some of the water in it, and thus of producing a more concentrated product. This evaporation is itself no inconsiderable portion of the business. It is conducted first in shallow tanks of lead, and finally in costly stills of platinum. When at length the acid in the platinum stills has attained the proper degree of concentration, it is drawn out by means of a siphon tube, and through a cooling tank of cold water, into the glass flasks called carboys, in which it makes its appearance in commerce.

Of course the account thus given is but a general sketch of this great industry. Associated with the apparatus and the processes here briefly described there are employed in actual working a multitude of other devices

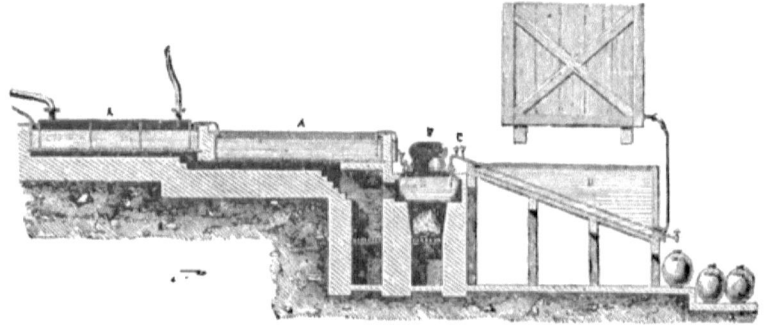

FIG. 47.—Section of apparatus used for concentrating sulphuric acid. *A, A*, laden pans in which the first evaporation is conducted; *B*, platinum retort in which the concentrating is finished.

and operations. Indeed it might be anticipated that the successful conduct of a business of such magnitude and complexity would draw upon the inventive resources of some of the best minds that have been brought to bear upon chemical industries.

READING REFERENCES.

Sulphur Industry in Sicily.
    Barbaglia, A.—Chem. News. xxxiv, 245; xxxv, 3, 28.
    Vincent, C.—Am. Chem. Journal. vi, 63.

Sulphur, Extraction of.
    Sestini, F.—Jour. of Chem. Soc. of London. xxviii, 335.

Sulphuric Acid.
    Affleck, J.—Chem. News. xxxvii, 167, 192, 207.
    Hasenclever, R.—Chem. News. xxxv, 48, 67, 88, 118, 183, 189, 214, 227.

CHAPTER XVIII.

BORON.

HE white substance called borax has long been known to exist as a solid deposit in the earth of many parts of the ancient East. But its uses have increased a thousand fold as the result of the modern discovery of new and far more abundant sources of it. Thus in the manufacture of porcelain and in other of the industrial arts, and as a remedial agency in medicine, borax has now come to be an important and truly useful substance to mankind.

The knowledge of its composition is referable to a very recent date; only in the present century its character as a true chemical salt was fully made out. Borax is now recognized as sodic borate, which usually exists in a form holding ten molecules of water of crystallization; accordingly the chemical formula is $Na_2B_4O_7 + 10\,H_2O$. From this it appears that, in addition to the well-known substances sodium and oxygen, borax contains a special and peculiar element called boron—a name evidently derived from borax. Again, being a salt, the substance must be viewed as containing an acid—or more properly speaking, the representative of an acid. That this is indeed the fact may be readily proved. If borax is dissolved in water in such a way as to form a concentrated solution, then, upon addition of chlorohydric acid, a solid substance separates out in pearly flakes; this upon subsequent examination is found to be an acid. This solid acid has received the name boric acid, and it may be represented by the formula H_3BO_3.

FIG. 49. — View among the Tuscan lagoni before the introduction of the borax industry.

(158)

Fig. 50. — View in Tuscany after the establishment of the borax industry.

(159)

Sources of Borax in Nature.

For a long time the only known source of borax was the natural crusts of this substance found principally in the ground in certain parts of Asia. At the present day, however, borax is obtained from Borax Lake, in California, in very large quantities. In fact the commercial needs of the United States for this substance, are readily supplied from borax found within its own borders. The most interesting and important step in connection with the preparation of borax dates back to about the year 1776, when the fact was made public that certain lagoons in Tuscany contained boric acid in their water. It was not until about the year 1828, however, that the manufacture of boric acid from this source was successful upon a large scale. In some of the Tuscan valleys there are volcanic crevices in the earth, called *suffioni*. From them steam escapes charged with certain compounds of boron. When this steam is brought in contact with water, boric acid is liberated in the water. The method of securing the acid is as follows: A ring of masonry is built in a suitable place and so as to include several suffioni. Sometimes new suffioni are artificially bored within this ring. Into the basin so produced, water from some convenient spring is conducted. The steam from the suffioni passing into the water, produces boric acid there. When the water is sufficiently charged, it is made to flow as a gentle cascade over a long series of shallow pans. The liquid readily evaporates from these pans for under them also, steam from suffioni is turned. It is indeed this last mentioned step in the manufacture, that became the turning point which has lead to its successful prosecution. The great cost of fuel for artificially evaporating the acid liquors, rendered unprofitable the earlier attempts to utilize this source of boric acid. A French gentleman, M. Larderel, suggested the use of steam from suffioni for the evaporation of the liquids produced, and the process was so successful that he quickly derived a colossal fortune from its employment. At the same time he enriched the territory that was previously not only desert and unproductive, but also was looked upon by the inhabitants with superstitious dread and as little better than the gate of the infernal regions. As a result of these inventions, a barren and unfrequented territory has been changed to a seat of thriving and beneficial industry. Finally, it is interesting to note that for his services in developing the boric acid industry M. Larderel was created Count of Monte-Cerboli by the Grand Duke of Tuscany.

READING REFERENCES.

Boric Acid, Manufacture of, etc.
Payen.—Annales de Chimie et de Physique. 3 Sér. i, 247; li, 322.
Dieulafait, L.—*loc. cit.* 5 Sér. xii, 318; xxv, 145.

Borax Lagoons of Tuscany.
Harper's Magazine. i, 397.

CHAPTER XIX.

NITROGEN.

NITROGEN is an important constituent of our atmospheric air of which it makes up about eighty per cent. The other twenty per cent, as has already been stated, is oxygen. In the air the nitrogen is found in the free or uncombined state, and we may reasonably suppose that it exists here to fulfil some important offices. Unquestionably one of these is that of diluting the oxygen, the energetic constituent of air, and lessening its activities—for nitrogen itself is extremely inert. From the part it performs in the atmosphere, nitrogen derives a considerable portion of the interest with which it is invested.

Discovery of Nitrogen.

Perhaps the first clearly defined recognition of nitrogen as a constituent of the air is referable to the genius of a wonderful man, who, in obscurity and with the most imperfect appliances, obtained an insight into the constitution of substances which has rarely been surpassed. Reference is here made to the Swedish, or rather Prussian, chemist Scheele, some of whose discoveries have been briefly adverted to in earlier chapters. It has already been stated that the distinct notion of a gas dates but little more than a hundred years back; and this statement is intended to call to mind that brilliant period in the history of chemistry when among others, Black in Scotland, Cavendish and Priestley in England, Lavoisier and his worthy associates in France, and finally, the sagacious Scheele in Sweden, were engaged in a generous rivalry in chemical studies, which made the close of the eighteenth century a period in the history of chemistry that will not be forgotten so long as the science itself shall endure. At this time unstinted effort was devoted, with ingenious but imperfect appliances, to the study of gases. Of course the atmospheric air, as the gas most vast in quantity, most accessible for experiment, most important in its relation to the economy of living nature, received its full share of attention. It was at this period that Dr. Rutherford, a professor in the University of Edinburgh, demonstrated that after living animals have breathed in a confined bulk or volume of air, there remains an inert and peculiar gas behind. And Priestley showed that after the burning of charcoal in a confined volume of air there remains a gaseous material equal to about four-fifths of the amount of original air used. But it was Scheele who first clearly pointed out that the air contained a second distinct constituent that fails to support combustion and animal respiration. And Lavoisier first proved this constituent to be an elementary substance and he gave to it the name *azote*, which it still retains in the French nomenclature of chemistry.

It is not forgotten that a critical examination of the history of human knowledge respecting the atmosphere reveals the fact that a wonderfully clear, even though incomplete, account of the functions of the active constituent of the air was printed as early as the year 1669, by an English physician named John Mayow.* This affords another illustration of the fact, recognized by all students of history, that often in the progress of knowledge, before the clear and full dawn there seems to be a twilight; at such a time, and before the darkness has been fully dispelled, there have been found here and there men gifted with supernatural vision who have been able to read the laws of nature long before acknowledged philosophers even had found light sufficient. And so the truths learned by Mayow, though clearly stated by him, failed of recognition until they were rediscovered a hundred years later. (See p. 121.)

Preparation of Nitrogen.

Nitrogen is usually prepared from the air by the withdrawal of oxygen from it. This withdrawal is effected by some substance which has a strong affinity for oxygen.

---
*KOPP, HERMANN: *Geschichte der Chemie.* Dritter Theil. s. 193.

Thus one method frequently resorted to for the preparation of nitrogen, is to burn phosphorus in air. Phosphorus is placed in a little crucible of porcelain and then floated upon a cork on the surface of water in a pneumatic trough. A bell-glass of air is now inverted over the phosphorus, after the latter has been set on fire. The phosphorus burns at the expense of the oxygen in the bell-glass. Thus the oxygen is little by little withdrawn and as a result the nitrogen is left.

Fig. 51.—Preparation of nitrogen from air, by absorbing the oxygen by burning phosphorus.

Another method for preparing nitrogen is based upon the same general principle. It is the following: Pass a current of dry air through a tube containing copper turnings heated to dull redness in a furnace. Under these circumstances the copper absorbs oxygen from the air, and leaves the nitrogen, which passes on to a receiver prepared for it.

Properties of Nitrogen.

Nitrogen prepared by these methods, or by any others, possesses the following characteristics:

It is a gas that is colorless, odorless and tasteless. It is not necessary to make any scientific demonstration of these facts, because with every breath of air drawn into the lungs of a human being a large quantity of nitrogen is inhaled, and it is easily perceived to be without odor or taste, while a glance of the eye into the atmosphere shows that, in moderate quantities at least, it is free from color. Up to a period dating but a few years back, nitrogen was spoken of as one of the permanent gases; and this word permanent was intended to convey the idea that it is not condensable to the liquid form. It is true that it was surmised that for every gas there must be a point of very low temperature and very great pressure at which the gas would assume the liquid form. Yet nitrogen, and two or three others, successfully resisted all such attempts to liquefy them until toward the close of the year 1878. Since that time, successful effort has been made to bring to a higher degree of perfection the appliances used for subjecting gases at once to intense cold and enormous pressure. With these, it is believed that small amounts of nitrogen have been liquefied. And it may even be said that there is now no permanent gas known, but that all gaseous substances may in fact be liquefied.*

As a simple and uncombined substance, nitrogen is characterized by extreme inactivity. It does not burn; it does not support combustion; it cannot be made to enter into chemical union with other substances, except by specially devised and circuitous processes.

While on the one hand inertness is the marked characteristic of the nitrogen, on the other hand this element is a constituent of a very large number of compounds. Moreover, these compounds are themselves often characterized by a high degree of activity. Of the last two declarations the first one seems to be inconsistent with the properties of nitrogen in its elemental form. The second one seems inconsistent, but less so when it is carefully considered. Thus the activity of the compounds of nitrogen is to a certain extent referable to their instability. The meaning of instability, as used here, is that the compounds are easily decomposed; and this is because the inert nitrogen readily lets go its hold upon the other elements. Whence it appears, that the activity of the compounds, in reality referable to the energetic action of the element or elements now loosed from the nitrogen, rather than the nitrogen itself.

In nature, nitrogen is found as a constituent in some very important compounds. Thus it seems to be an essential element of some of the principal animal matters, such as muscular fibre and the material of the brain. Again, it is a constituent of ammonia gas and also of a multitude of compounds derived from it. Now these compounds are members of a group of substances which serve as most valuable kinds of food for living plants. So it may be said that both living animals and plants seem to be in a peculiar way dependent upon nitrogen or nitrogenous matters.

---

*SCHUTZENBERGER, PAUL: Traité de Chimie Générale, Paris, 1880, i, 30.

Compound of Nitrogen with Hydrogen.

Under favorable circumstances, nitrogen and hydrogen combine to form the stable, interesting and important compound called ammonia gas and having the formula NH_3.

While this gas may be produced by the direct union of the constituents—that is when a mixture of nitrogen gas with hydrogen gas has an electric discharge slowly passed through it—this is not a common mode of procedure. Ammonia gas is oftener produced by a natural or artificial decomposition of certain substances that contain nitrogen and hydrogen among their constituents. As it has already been stated that many animal matters contain nitrogen and hydrogen, it follows that animal matters when decomposed, afford ammonia gas; and so they do, in fact, whether the decomposition is in the course of their natural decay, or whether it is conducted artificially, as for example when *animal* matters are heated in closed vessels to the point of decomposition. Indeed ammonia gas and its important commercial compounds were formerly produced in this last mentioned manner.

Ammonia gas—or some compound of it—is also formed, as may be readily imagined from what has already been said, from decomposition of *vegetable* matters containing nitrogen. It is a fact that at the present day the principal supplies of ammonia gas and its compounds for the uses of commerce and the arts come from such a source, that is from the artificial decomposition of bituminous coal. It is true that in the ordinary sense coal is not vegetable matter. But careful examination of it, shows that it is very directly derived from the vegetation of ancient forests. The vegetable matter has been packed away in the earth and has been subjected to water, heat and pressure under such conditions that these agencies have changed it to the form in which we find it. Now the coal-gas industry of the present day is so conducted as to decompose coal and collect many of the products of its decomposition. One of these products is ammonia gas. To the decomposition of coal, therefore, the business world at present looks for its supply of ammonia gas and the many compounds derived from it.

The name ammonia gas, indicates that it ordinarily exists in the aeriform condition. It has a very pungent odor, well-known as that evolved from smelling-salts. It dissolves in water with very great facility and in very large quantities. It has a strong tendency to combine with acids. This last fact may be easily illustrated by simple experiments within the reach of almost any one.

Experiment with Ammonia.

Provide two wine-glasses or two shallow vessels of any sort. Into one of them pour the liquid known as spirits of hartshorn, and called by the chemist ammonic hydrate. Into the other pour some concentrated chlorohydric acid. Abundant white clouds will quickly form above the vessels and between them. These clouds are composed of minute particles of a solid, called by the chemist ammonic chloride and expressed by the formula NH_4Cl. The reason for their formation is this: from the spirits of hartshorn escapes ammonia gas (NH_3); from the acid there constantly escapes chlorohydric gas (HCl); the two gases meeting in the atmosphere combine with energy, and form the smoky product referred to.

The chemical change is represented by the following equation:

NH_3	+	HCl	=	NH_4Cl
One molecule of		One molecule of		One molecule of
Ammonia-gas,		Chlorohydric acid,		Ammonic chloride,
17		$36\frac{1}{2}$		$53\frac{1}{2}$
parts by weight.		parts by weight.		parts by weight.
$53\frac{1}{2}$				$53\frac{1}{2}$

The ammonic chloride thus produced is an article of commerce, well-known under the name *sal ammoniac*. As has been said, it is a solid and it belongs to the class of substances designated by chemists as salts. In fact one of the most striking characteristics of ammonia gas is its power to produce salts by union with acids. Here is a list of three well-known salts of this sort:

With Chlorohydric acid, HCl   it produces Ammonic chloride, NH_4Cl
"   Nitric acid, HNO_3 it produces Ammonic nitrate, NH_4NO_3
"   Sulphuric acid, H_2SO_4 it produces Ammonic sulphate, $(NH_4)_2SO_4$

Compounds of Nitrogen and Oxygen.

Nitrogen and oxygen ordinarily manifest scarcely any affinity for each other. There are conditions however under which they unite; and more over they unite in different proportions so as to form at least five different compounds. These may be presented in the form of the following striking series:

Nitrogen protoxide (called laughing-gas,) N_2O.
Nitrogen dioxide, N_2O_2 (or NO).
Nitrogen trioxide or nitrous anhydride, N_2O_3.
Nitrogen tetroxide (brown fumes,) N_2O_4 (or NO_2).
Nitrogen pentoxide or nitric anhydride, N_2O_5.

Of these compounds, unquestionably the most important is nitric anhydride—and this not on account of itself, for it is very rarely produced either in the arts or in the investigator's laboratory. Its importance is referable to the fact that added to water, it produces nitric acid.

This chemical change is represented by the following equation:

$$N_2O_5 + H_2O = 2HNO_3$$

One molecule of Nitric anhydride, 108 parts by weight.	One molecule of Water, 18 parts by weight.	Two molecules of Nitric acid, 126 parts by weight.
126		126

Nitric Acid.

This acid has been referred to in another place as one of three principal acids of commerce. Certain of its most striking properties may be displayed in an easy and interesting manner by any one. For this purpose the following experiments are suggested:

First experiment.—Nitric acid turns quill yellow.

Place a few fragments of white quill in a test-tube. Add a few drops of nitric acid and then some water. Now warm the mixture. The quill will be found to acquire a yellow color. Fill the tube with cold water in order both to dilute the acid and to cool it. Pour away the liquid, and wash the quill in water. The yellow color will be found to be permanent. Many other animal matters are turned to a permanent yellow color by nitric acid.

Second experiment. — Nitric acid attacks copper with violence. There is liberated by the process a gas called nitrogen dioxide (N_2O_2), which is colorless but which becomes brown upon exposure to the atmospheric air. The chemical change gives rise to a solution sometimes green and sometimes blue, according to circumstances.

Place in a test-tube a small piece of metallic copper in the form of either wire or foil. Add some nitric acid to the copper. Then warm it gently until the copper disappears. The brown fumes will be recognized. The colored solution of cupric nitrate, $Cu(NO_3)_2$ should also be noticed.

Third experiment. — Nitric acid attacks zinc with great violence.

Try another experiment quite similar to that just described, only employ zinc in place of copper. Brown fumes are evolved, and a colorless solution is produced containing zinc nitrate, $Zn(NO_3)_2$.

Fourth experiment. — Nitric acid attacks iron with violence.

Try another experiment, quite similar to the second and third, only employ iron instead of the other metals mentioned. The fine iron wire used by florists is suitable for this purpose. The same brown fumes are evolved. A metallic nitrate is also produced; it is called ferric nitrate and its formula is $Fe_2(NO_3)_6$. The solution is yellow, or but slightly colored.

Fifth experiment. — Nitric acid dissolves a nickel coin.

An experiment similar to those already detailed may be tried upon a nickel coin; but it is not necessary to entirely dissolve the coin. After the acid has acted for a few moments, water may be poured into the tube so as to dilute the acid, and at the same time to cool it. Then the liquid may be poured away and the coin withdrawn. In addition to the brown fumes evolved, the feature most noticeable is the decided green color of the solution. This is referable, to a considerable degree at any rate, to the nickel present. Nickel imparts a green color to most of its solutions.

These experiments suggest that nitric acid has a marked influence upon the metals. This is in fact one of its prominent characteristics; and it is largely used in the arts for the purpose of dissolving metals.

CHAPTER XX.

THE ATMOSPHERE.

THE atmosphere or the air of our globe is the vast ocean of gaseous matter at the bottom of which human beings, as well as other land animals, dwell. While it is so thin that a vessel full of it is spoken of in ordinary language as being empty, it yet possesses a reality which it often displays in a very serious manner. Its presence is made gently evident to human beings by the moderate resistance it offers to them when they are in motion; but when itself is in motion with the force of the hurricane or tornado, no solid matters can stand in its path. Heavy railroad trains and massive buildings are hurled from their positions and turned into miserable masses of wreckage, while even strongly rooted forests are swept out of place by its vigorous breath. The terribly destructive power of air at one moment and and its mild and subtle efficiency at another are very suggestive of the wonderful adjustment of the forces residing in it. It is by the restrained action of these forces that the atmospheric air is so admirably fitted to perform its varied functions in connection with animal life. At the same time it is so unobtrusive in its workings that its very existence is at first scarcely noted. When at rest, it peacefully wraps the earth about as in a gossamer veil, but when in angry agitation it scourges country and city alike as with a whip of gigantic cables.

The height to which the atmospheric air extends above the earth is not exactly known. But carefully devised experiments have shown that going upward, its compactness or density diminishes very rapidly. Indeed calculations based upon exact experiment show that at a height of forty miles, or thereabouts, from the surface of the earth, the air is so highly rarified that practically it there comes to an end. In other words, at this height a given bulk of space contains no more air than exists in the so-called vacuum produced by a superior air-pump.

Weight of Air.

Notwithstanding the extreme tenuity of the gaseous medium in which we live, it is capable of buoying up on its wings a multitude of living beings of vast aggregate weight. It is firm enough to support the millions of birds that sail in it, and the myriad of millions of insects who yet more freely navigate it in search of food and warmth, and in answer to the various needs of their existence.

One of the most striking evidences of the fact that air is indeed a material substance is very easily discovered by showing that it possesses weight. Thus suppose a properly constructed glass globe is almost entirely emptied of air by the action of an efficient air-pump. Suppose then that the globe is weighed. Next if it be connected with a bell-glass containing ordinary atmospheric air over a pneumatic trough, it may be readily seen that the air leaves the bell-glass in order to pass into the globe. If this globe is now weighed again, it is found to manifest a decided increase of weight. This increase is due to the air it has received. By such means it may be easily shown that a cubic yard of air weighs not far from two pounds.

Composition of Air.

The principal constituents of air are the two gases oxygen and nitrogen; and of these the oxygen makes up about one-fifth and the nitrogen about four-fifths of the whole. In addition to these principal substances, however, certain others are always present, of which may be specified vapor of water, carbon dioxide and ammonia gas; while more minute quantities of a vast multitude of other gaseous substances find a reservoir in the air. It is an unquestioned fact that the atmosphere is likewise charged most of the time with still more minute quantities of solid dust materials of various kinds. An example is found in the common salt, blown up into the the atmosphere from the ruffled surface of the oceans. Now the oceans are spread over fully three-fourths of the earth's surface, and the winds, blowing upon the crested waves, not only diffuse the salt over the oceans themselves but also carry it far inland; accordingly spectrum analysis reveals the presence of salt in almost all atmospheric air.

Just as the rivers of water, flow to the ocean and bear along to it debris of every kind—pulverized rock and earthy materials and other washings from the soil, leaves of forests, impure products of civilization thrown in from houses and manufacturing establishments—and all these materials make their relatively minute contributions to the impurities in the great ocean itself, so it is with the atmospheric ocean. Thousands of millions of living animals pour out, with every breath from the lungs, materials exhaled from their bodies. And so wherever fuel is burned, or wherever manufacturing establishments liberate gases or vapors, or even finely pulverized solids, these are cast forth from the mouths of their reeking chimneys; and they all flow into the great aerial sea. So then it is no unexpected circumstance that the air should be a reservoir in which, in minute quantity, is likely to exist every gaseous substance produced.

Offices of the Several Constituents of the Air.

The oxygen of the air is its most active constituent. This is the substance that has already been described as essential for all ordinary combustion and all animal respiration. By a great variety of characteristics it is well fitted for these important offices.

The chief duty of the nitrogen appears to be to dilute the oxygen and moderate the excessive activities that would be manifested if the atmosphere consisted entirely of the active gas. Since iron and other metals burn in pure oxygen, it is plain that in an atmosphere of oxygen—containing no moderating gas like nitrogen—a fire once kindled in a stove would not confine itself to its proper fuel, but would soon spread to the metal of the stove itself, and so initiate conflagrations that could hardly be restrained.

The *moisture* in the air adds a number of wonderful and serviceable characteristics to it. Thus it helps to retain the heat received from the sun and so materially contributes to the sustenance of animal and vegetable life. The heat of the sun penetrates our atmospheric coverlet with great readiness and this heat is received by the surface of the earth and thence is imparted to the layer of air immediately upon it. Now the moisture contained in the atmosphere—and in principal quantity in the portions of air closest to the earth—is one of the chief agencies that prevent the immediate escape of that heat that the solid earth has secured from the sun's rays. And it is in the warm layer of air so produced that animals and plants chiefly flourish. Ascend a mountain's side and a height is soon reached at which eternal snow and cold prevail, where animal life cannot penetrate and even the lowest forms of vegetable life can hardly make their residence. What has thus far been said points out a valuable office of watery vapor and one that is entirely in addition to that which this same material performs as it floats in the clouds, ready to fall as beneficent showers and then to proceed to the other steps in the progress of that useful circulation which it performs as a liquid. But it may not be out of place to mention here that the aqueous vapor in the atmosphere appears to serve in another way for man's pleasure, even though in this particular no utility can be claimed. Thus the glories of sunrise and sunset, which have delighted intelligent beings for so many ages, are paintings upon the drapery of the firmament which the pencil of light has been enabled to produce through the medium of the refractive power of those gathering drops of water which float about in various forms and combinations in the morning or the evening sky.

It has already been more than once declared that *carbon dioxide* is poured out into the atmosphere by all the ordinary processes of combustion. This is not only true of combustions such as those of coal and wood and similar highly carbonaceous materials; it applies with equal force to the animal body itself, which has been properly likened to a furnace. The air taken into the lungs at each breath inspired, supports during life a continual combustion, by reason of which, minute fragments of the animal tissue are burned in all parts of the system. One of the products of this burning is carbon dioxide, which is carried to the lungs, thence to be exhaled as a waste product into the atmosphere. It might at first be expected that this carbon dioxide would accumulate and form a constantly increasing proportion of the air. But it is one of the proper foods of vegetable life; for nature has wonderfully provided that plants should thrive by the absorption or inhalation of this particular gas. And so all the leaves in the forest are continually cleansing the air of that carbon dioxide that living animals have cast aside as a useless thing. And by a magnificent alchemy, the result of a wonderful and beneficent plan, they turn this waste matter of the animal frame into food for themselves, and they cast out into the air as *their* refuse that oxygen gas which living animals demand. So then the two forms of living beings exist in a harmonious partnership by reason of which each one is benefited.

An example similar to that just given with respect to carbon dioxide is found in *ammonia gas*. This substance is one of the commonest products of the decay and decomposition of animal matters. Wherever animal waste is deposited upon the surface of the earth it quickly evolves ammonia gas. This gas diffuses itself through the atmosphere under the influence of conditions whereby it may perform an important service; for it is always extremely soluble in water. And so as soon as rain is condensed, whether in a gentle shower or in abundant torrents, each drop in passing through the air gathers ammonia and carries it down to the earth. Again, ammonia is one of the chief foods for plants. And so the rain drops, charged with such ammonia as they have been able to collect, bear it to the rootlets in the soil, as a valuable and important food, and one which has been proved to have a most stimulating influence upon their growth.

There is not opportunity here for a discussion of the offices and the interplay of the other substances existing in atmospheric air; for they are more local in their effects and more difficult to trace and to describe.

The Air is not a Chemical Compound.

The importance of the atmosphere and its great abundance have naturally led to most thorough scientific scrutiny of it. Thus the amounts of its principal constituents have been studied with extreme care. One result has been that the principal constituents—the oxygen and the nitrogen—have been found to exist in air in proportions singularly constant in amount. This fact indeed has suggested to some chemists the impression that air is a true chemical compound. This latter suggestion appears not to be sustained by the most rigid examinations that have been made. In fact they give ample support to the opinion already declared—that the air consists of a mass of merely mingled gases and that these gases are uniformly maintained in their proper proportional amounts by the beautiful interaction of the physical and chemical properties with which they are endowed.

Fitness of Atmospheric Air for its Uses.

The statements already presented must have suggested to the reader that the atmospheric air fulfils its uses in nature much as any contrivance carefully devised by an intelligent framer would accomplish the work for which it was planned. Beside those chemical adaptations which have been the principal grounds upon which this line of thought has been supported here, there are others which may be briefly suggested.

By reason of the *mobility* of air, as well as its tendencies to expansion by heat, our atmosphere is necessarily in a state of most intricate ebbing and flowing. One prominent effect of the motion thus set up is to cause a transfer of warm air, and so a distribution of heat, from more favored portions of the globe to the others.

This same result is also more completely attained by the influence of the *specific heat* of air. Atmospheric air has remarkable power, in which it resembles to some extent water: to take up a very large amount of *heat* with but a slight rise in *temperature*; similarly a slight fall of temperature is associated with a large evolution of heat. By reason of these properties, air, like water, has an exceptional storage power for heat. This contributes largely to the equalization of climates.

The *elasticity* of the atmosphere permits it to become a useful servant of man in the transmission of sound. Thus human beings—and with less distinctness most of the living creatures of the lower orders—communicate their thoughts by means of spoken words through that line of atmospheric air extending from them to their hearer or hearers.

Again the characteristics of the atmosphere are such that it *diffuses* sunlight. By this is meant that in air, sunlight does not confine itself to those strictly straight lines which it follows in empty spaces. So then this property of air mitigates the blackness of shadows and, for example, he who walks into a shady lane does not plunge into absolute darkness, as he might if we were deprived of this beneficial diffusing influence of the atmospheric air.

The two considerations last adduced contribute much towards making the earth a cheerful home for human beings; for they aid materially in the distribution of intelligible ideas. Moreover those properties of air by virtue of which its undulating waves makes music possible, and further those which permit the flight of light, and so allow of the existence of the graphic arts, certainly make no mean contributions to the happiness of man; and thus they help to furnish the earth as his place of residence.

Fig. 53.—Display of fireworks on the Seine, Paris.

CHAPTER XXI.

EXPLOSIVES.

THE principal explosives owe their activity, to a very large degree, to the presence of nitrogen in them; thus they may properly be discussed in connection with that element.

The explosives of chief importance are four in number: gunpowder, the fulminates, gun cotton, nitroglycerine. While these substances suggest at once the war-like uses to which they are put, it must not be forgotten that they have also important applications in the arts of peace. Thus enormous quantities of gunpowder and nitroglycerine are used in blasting operations for purposes like the removal of rock preparatory to laying foundations for large buildings, as well as in excavations for railway cuttings and in the boring of tunnels; also in the getting of building stone from quarries, the tearing of ore out of mineral bearing veins in mining operations; and for loosening coal in coal pits. Large quantities are likewise employed in pyrotechnics. It must not be forgotten that fireworks are not only for purposes of night illuminations and for public gratification in times of popular rejoicings; they are also employed to a considerable extent for such useful purposes as night signalling on vessels at sea.

Gunpowder.

Of the various explosives mentioned, gunpowder is the oldest. While the invention of this substance has often been referred to Roger Bacon, the celebrated English friar who died about 1292, it is now conceded that though Bacon evidently knew the composition of it, the original invention dates far earlier than his times. There seems foundation for the belief that it is as old as a thousand years, while its use by artillery at the battle of Crécy shows its employment in warfare for over five hundred years. Bacon's power of independent thought, placed him so far in advance of the century in which he lived that he became an object of persecution, but he is at present ranked as one of the prominent figures of history. In his works Bacon refers to a substance that seems to correspond to gunpowder, and in terms that suggest that he considered it as a material of not uncommon knowledge in his day.

The principal constituents of gunpowder are three: potassic nitrate, charcoal and sulphur. The chemical action between potassic nitrate and charcoal in gunpowder may be better understood after a simple experiment, which any one can try. The experiment

FIG. 52.—Roger Bacon, born near Ilchester, about 1214, died probably at Oxford, in 1292.

referred to is as follows: take a large piece of charcoal; heat it over a spirit lamp or gas lamp until certain portions of it take fire so as to burn with a slight glow; next sprinkle very carefully a small amount of powdered potassic nitrate—also called both saltpetre and nitre—upon it. A burning, something like that of gunpowder, only less violent, results. The potassic nitrate has the formula KNO_3. When it falls upon the glowing coal a portion of the oxygen leaves the other constituents and accomplishes thereby a true combustion of the carbon. One important factor in the operation is the element nitrogen; owing to the general inertness of nitrogen it easily allows the escape of other elements combined with it. So in case of the experiment just suggested, the combustion of the charcoal is referable to oxygen liberated by reason of the feeble affinity of one of the other constituents of the potassic nitrate—that is, the nitrogen. Thus far the only thing particularly suggested is the combustion that takes place; another point of importance may be mentioned in this connection. If finely powdered charcoal and potassic nitrate are thoroughly intermingled and then set on fire in a closed vessel, a large amount of a gas, that is, carbon dioxide, will be generated by the combustion; and this gas may burst the vessel unless it is a very strong one. If, however, the vessel has an opening supplied with a cork or plug, this stopper will be violently driven out by reason of the explosive force of the carbon dioxide generated. So in the preparation of gunpowder, potassic nitrate, charcoal and the third substance sulphur, are finely pulverized and carefully intermingled. Thus they are brought to a state of thorough diffusion and intimate contact. The offices of carbon and potassic nitrate have been already explained. The office of the sulphur is principally to combine with the potassium of the potassic nitrate, producing as a result a somewhat larger evolution of gas. At all events, when gunpowder is consumed, two important results are afforded. As already intimated, the first is the sudden liberation of a very large amount of gas—carbon dioxide. The second is that this gas is generated by a process of true combustion attended with great heat, the latter contributing largely to the explosive force by reason of the great expansion of the gaseous products, effected by the heating.

There are several different kinds of gunpowder, but they all consist essentially of the constituents mentioned. Their differences are either in the proportions of the constituents used or in the size of the granules in which the powder is formed. Thus for some war purposes it is requisite that the powder should burn very rapidly, while in others it is required to burn slowly. For the purpose of regulating the rate of combustion, the grains are made of various sizes. The smaller sizes burn more quickly, while those of larger dimensions as well as those more strongly compressed, burn more slowly.

While the exact chemical changes which take place when gunpowder burns, are too complicated to admit of discussion here, they are in the main those just explained.

Fireworks.

Gunpowder affords the basis of the pyrotechnic art. It is employed also with the distinct intention of utilizing both of those prime properties already referred to. That is to say, by reason of its explosive force, gunpowder produces the various forms of *motion* requisite in fireworks. By reason of the intense heat afforded by its combustion, the various kinds of *light* are producible. The truly marvellous effects obtained by the skilled pyrotechnist involve the use of a great multitude of substances and also an ingenious mechanical combination of them. The effects he must produce require brilliant light in various qualities and also upon occasions loud reports, as the bursting of bombs and the like. So many forms and combinations of fireworks are possible that no enumeration can be made here; moreover, their infinite capabilities depend upon the inventive resources and skill of the maker. In a brief description, the rocket may be taken as the type of fireworks. It is often of most ingenious construction. Thus it may be provided with many chambers, one connecting with another by proper passages. In these passages are placed fuses so that the fire shall run from one chamber to another in proper order. Of course the main barrel contains a quickly burning gunpowder. This is for the purpose of producing the ascent. It is well known that a pistol, a rifle or a cannon, always experiences a strong recoil when fired. So does a rocket; but the rocket is so constructed that the recoil is the chief factor in its first discharge. That is, if the rocket is compared to a cannon, the discharge is downward and the recoil upward, so that in fact the ascent of the rocket is due to what may be called an exceedingly powerful recoil. When the rocket is high in air, the fuse connected with its principal barrel lights its subordinate chambers, and these then exploding distribute into the sky the brilliant masses of stars or other graceful pieces originally intended. The loud reports that take place at such times are due to portions of violently explosive substances within certain chambers;

while the party-colored lights produced are referable to the burning of substances which have been carefully selected for the purpose. Thus the pyrotechnist has recourse to mixtures of gunpowder and various other chemical substances to produce colored fire. Finely powdered charcoal or lamp-black give rise to a red fire; so also do most of the salts of strontium. Common salt or powdered rosin give rise to yellow fire. Copper filings and certain salts of copper produce greenish hues; so do salts of barium. Zinc filings and chloride of copper, and certain others, produce blue shades. Saltpeter in considerable quantity affords a delicate pink; while iron filings and steel filings produce scintillations of great brilliancy.

Fulminates.

The fulminates are substances that are so extremely unstable in chemical character, that they require but a very slight mechanical blow to decompose them. Two fulminates in particular may be mentioned: fulminate of mercury and fulminate of silver. They are both viewed as salts of a peculiar complex acid called fulminic acid. This acid is a compound of carbon, hydrogen, oxygen and nitrogen. When silver or mercury takes the place of the hydrogen in fulminic acid, the dangerous salts just mentioned are obtained. Fulminating mercury is the one of chief use. It is employed in percussion caps. A drop of gum is put in the inside of the cap, then the exact amount of fulminate in the form of a powder is allowed to fall into the gum; finally the whole is allowed to harden. When the cap is used, a violent blow from the hammer of the gun or pistol gives rise to the explosion of the fulminate, and this communicates to the gunpowder of the cartridge to be fired. Fulminating silver is too dangerous for use in percussion caps, but it is employed in certain explosive toys like torpedoes.

Gun-Cotton.

Gun-cotton is a chemical modification of the ordinary cotton fibre. This fibre when purified by chemical washings consists entirely of the substance called cellulose. It is not different from certain other vegetable fibres. It has the formula:

$$C_6H_5O_5H_5$$

which may also be represented as follows:

$$C_6H_5O_5 \begin{cases} H \\ H \\ H \\ H \\ H \end{cases}$$

When clean cotton is acted upon by strong nitric acid it undergoes the wonderful chemical change to gun-cotton. Without material alteration in its physical appearance there has been a chemical substitution by reason of which a nitrogen compound has been introduced into the chemical molecule, as a substitute in place of certain of the hydrogen atoms originally present. Thus the formula of gun-cotton may be represented as follows:

$$C_6H_5O_5 \begin{cases} H \\ H \\ NO_2 \\ NO_2 \\ NO_2 \end{cases}$$

A comparison of this formula with the one given for pure cotton shows that three atoms of hydrogen in the cotton have been replaced in the gun-cotton by three molecules of the compound radicle NO_2. On this account gun-cotton is often spoken of as trinitrocellulose. By reason of this chemical substitution the cotton changes as if by magic from the simple, safe material ordinarily known, to one of the most dangerous of explosives. Thus Mr. Abel, the chemist to the English War Department, who has made a series of most careful studies of gun-cotton with reference to its use for war purposes, finds the explosive power of gun-cotton to be more than fifty times that of gunpowder of equal weight. One of the greatest objections to the use of gun-cotton is found in the fact that upon keeping, it

Fig. 54.—Establishment for manufacture of dynamite, near Turin.

undergoes of itself a steady decomposition resulting ultimately in dangerous explosions. This fact appears to be likely to prevent the substance coming into general use.

Nitroglycerine.

Glycerine—produced at present in enormous quantities from fats and oils—is well known as a sweetish, oily and harmless substance. Glycerine is composed of carbon, hydrogen and oxygen in proportions but slightly different from those in cotton. Thus its formula is

$$C_3H_5O_3H_3.$$

If this bland and simple material is subjected to the action of concentrated nitric acid, it undergoes a change very similar to that recognized in the case of cotton and just described. It then produces a compound called trinitroglycerine which, while it ranks as one of the most powerful and useful explosives, is also associated with a long list of horrible disasters produced by accidental, or in some cases intentional, explosion of it.

Nitroglycerine is itself an oily material and was at first considerably used in that form. The terrible accidents from transportation of the article have given rise to the adoption of two means for lessening the risks attending it. The first is the manufacture of the substance in suitable localities —that is near to great public works in which it is to be employed. And again the factories are so arranged that the operation of the manufacture shall be conducted in small buildings surrounded by earthworks sufficient to localize any explosion that might unhappily occur.

At the manufactory of explosives at Ardeer on the Scotch coast, about fifty miles from Glasgow, a most ingenious additional precaution is taken. Here each laborer, as he enters the works in the morning, passes into a cottage to change his dress. He dons a uniform of a special and distinctive color—it may be scarlet, or bright blue or white or gray, according to the department in which he is employed. Thus the policemen who are constantly on duty can detect at once any employé who strays into a

Fig. 57.—Torpedo-boat attacking a large war vessel.

department to which he does not belong and where his lack of acquaintance with the processes might lead to a terrible accident.

Another special device is the invention of Albert Nobel, who has been noted as the principal person by whose efforts nitroglycerine has been introduced into the important uses which it finds at the present day. This is the absorption of the liquid nitroglycerine in some spongy material such as will serve as a safe and proper vehicle for the explosive. The substance thus employed is a kind of fine siliceous earth called diatomaceous earth, also infusorial earth. This is a mineral material found in various parts of the world in somewhat abundant deposits. Upon examination by the microscope it is found to be composed of the mineral skeletons of microscopic organisms. The minute cellular texture which this substance affords seems to be admirably fitted to imbibe the liquid nitroglycerine, and assist in packing it in proper cartridges. The explosive produced by the combination is the one commonly known as dynamite.

Fig. 56. — Dynamite exploder.

A peculiarity of nitroglycerine and dynamite is that they cannot be fired in the ordinary fashion. That is, if a lighted match is brought to them they may take fire and burn with perfect quietness. For their *explosion* they demand some kind of violent blow. For this reason their cartridges have to be provided with special exploders. These are small cases of gunpowder or perhaps fulminating materials, which may be set on fire by means of a powder fuse or an electric current; their explosion within the nitroglycerine mass determines a violent shock to the latter. It is the concussion thus produced that is the appropriate means of exploding the nitroglycerine or dynamite cartridges.

While the sad accidents with these materials have horrified the whole world by their sudden and disastrous results it is too often forgotten that their gigantic forces are day by day safely and quietly contributing to the execution of great public works all over the earth. Thus in the great rock tunnels of Mont-Cenis and St. Gothard, which pierce the Alps, nitroglycerine and dynamite have done the work of armies of men. In the St. Gothard tunnel more than two million pounds of dynamite have been employed, and it has proved wonderfully effective in advancing most arduous subterranean work. Unquestionably the principal use of this explosive, as well as others, is in the labors of peace. Still, nitroglycerine and dynamite have come into great prominence by reason of their use in naval warfare. Torpedoes of a great variety of forms are now constructed so that a quick moving launch may steam up to a large ship of war, place close to her side one of these dangerous contrivances and then quickly withdraw in time to avoid the effects of the explosion which involves the great vessel in devastating ruins. Torpedoes charged with nitroglycerine or dynamite, are also used for the defence of harbors, being sometimes placed in such a way that an enemy's ship, in crossing the line formed by the torpedoes, shall by that act explode one or more of them and produce her own destruction.

READING REFERENCES.

Explosive Agents.
    Abel, F. A.—Jour. of Chem. Soc. of London. xxiii, 41, xxvii, 536.
    —Chem. News.—xxxix, 165, 187, 198, 208.

Explosives, a New Class of
    Sprengel, H.—Jour. of Chem. Soc. of London xxvi, 796.

Explosives, Force of
    Berthelot.—Annales de Chimie et de Physique. 4 Sér. xxiii, 223.

Explosives, in Blasting.
    Scribner's Monthly. iii, 33.

Greek Fire, (so called.)
    Lalanne, L.—Annales de Chimie et de Physique. 3 Sér. iv, 433.

Gun-cotton, Manufacture and Composition of
    Abel, F. A.—Journal of Chem. Soc. of London. xx, 311, 505.

Gunpowder, Chemical Theory of
    Debus, H.—Chemical News. xlv, 91.

CHAPTER XXII.

PHOSPHORUS.

HOSPHORUS is a most interesting chemical element. This is because of its exceptional chemical properties, the very important part it plays in the chemistry of animal and vegetable life, and its employment in the friction match, one of the most convenient and useful articles of human invention.

Phosphorus appears to have been first prepared in the year 1669 by a Hamburg merchant named Brandt who became fascinated with the study of alchemy and pursued his experiments with the view of repairing his broken fortunes by the discovery of the philosopher's stone. The happy discovery of phosphorus, while it did not enrich him, at least preserved his name in the annals of chemistry. Brandt produced it by a laborious process from certain animal matters. Notwithstanding the remarkable properties of the substance and the extraordinarily useful purposes to which modern scientific knowledge has applied it and its compounds, phosphorus remained the merest toy for more than a hundred years. In 1771 Scheele revealed to the world the fact that it may be prepared from bone-ashes, that is from burnt bone, and this has ever since been found to be its most convenient source.

The name phosphorus is derived from two Greek words ($φῶς$ *phos*, light, and, $φέρω$ *phero*, I bear) which suggest one of its marked properties, namely its power of continually affording light even though not set on fire after the manner of ordinary illuminating materials. It is true the light is feeble and chiefly noticeable in the dark. It is the same, in fact, as that yielded in the dark by an ordinary friction match when it is gently rubbed, but has not yet taken fire. This light, however, is the product of a true combustion, only of a very slow one; and again this burning of phosphorus is initiated by heat, (though only a very moderate amount is required for it). Of course for phosphorus much less heat is demanded than to set on fire our ordinary combustibles.

Phosphorus, though very widely distributed in nature, is never found free or uncombined. This fact is distinctly referable to the ease with which the substance combines with oxygen; if it were found free at any point on the surface of the earth, where it suffered exposure to atmospheric air, it would of course quickly enter into combination with oxygen.

Phosphorus exists occasionally in the earth in the state of combination in very hard rocky masses, of which the mineral known as apatite—composed mainly of calcic phosphate—is a good example. It is also present in small quantities in almost all soils; and in minute quantities in most natural waters, like river-water and sea-water.

One of the most familiar substances containing phosphorus is the bony skeleton of the higher animals. Here also it exists as calcic phosphate. It exists also in the brain, though in a form of chemical combination not easily stated.

Further, it is a constituent of various portions of the vegetable structure, especially of seeds.

The statements in the last two paragraphs have been presented with the express purpose of calling attention to the important offices of phosphorus in connection with animal and vegetable life. Thus exact experiments have shown that plants cannot flourish in soils barren of phosphates, and that the mere addition of almost any soluble phosphate to an arid soil promptly stimulates the plant living upon it, into more luxuriant growth. These facts have led to the introduction into commerce of artificial fertilizers containing soluble phosphates as their principal ingredients; and the manufacture of such fertilizers has continually expanded, until now it is conducted by the principal commercial nations on a truly gigantic scale. For the purpose of this manufacture, *bones* are particularly favorable because of their porosity. In fact the surface of the world is ransacked to supply this raw material. Thus from the deserts of Africa, bones are conveyed as far as England to be manufactured into fertilizers; and so from the great western plains of the United States, bones are brought to the eastern centres for a like use.

The agricultural demand for phosphates of some sort has become so imperious that even apatite is now largely used, notwithstanding the difficulties that its exceedingly hard and compact structure place in the way of the manufacturer.

From the plant, phosphorus finds its way in the form of food into the animal system. The living animal appeciates this essential ingredient, carefully selects it out from the food, and stores it up both in its brain and in its bony framework. This framework is exceedingly important as giving the requisite rigidity to the whole structure, and the proper support for the action of the various muscles.

Phosphorus itself is prepared by a process too complicated for the ordinary amateur chemist to repeat; indeed its preparation, even on the large scale, presents serious difficulties. These are associated with the great combustibility of the substance, which makes necessary extraordinary precautions against fire. Again, laborers in phosphorus works are subject to a painful and incurable disease called *phosphorus necrosis*, which has a peculiar and destructive effect upon the bones of the jaw. Finally, the chemical changes involved give rise to such difficulties and complexities as force the manufacturer to unusual watchfulness. In fact it has been recently stated that there are scarcely more than two factories for phosphorus manufacture in the world—one in England and one in France.

The element phosphorus, as ordinarily seen, has much the appearance of wax. It has a white or amber color, and is translucent; it may be cut with a knife much as wax cuts. It is ordinarily sold in the form of cylinders of about half-an-inch in diameter. It is necessary to keep it in vessels of water, for as already stated, if exposed to the air it would oxidize. This oxidation, at first slow, increases in vigor from the heat afforded by the earlier stages. After a short exposure to air, portions of phosphorus spontaneously burst into flame. Evidently then, phosphorus should not be handled except under water. Cases are recorded of severe and even fatal burns—the result of handling phosphorus in the air.

We may with propriety call attention here to another peculiarity of phosphorus, which constitutes one of the remarkable features of this interesting element. About thirty years ago, a Vienna chemist discovered that when phosphorus is heated for a considerable length of time, under conditions such that no gas is present which can act chemically upon it, it undergoes a marked change in its properties. Thus its color turns to red, and, strange to say, it loses altogether that ready combustibility which is the most striking characteristic of ordinary phosphorus. It may seem incredible that any such change could in fact occur. But this red phosphorus has become an article of considerable importance in commerce, and it is a well-established fact that ordinary phosphorus may be turned into this modification without any gain or loss of weight, and that, on the other hand, this red phosphorus may be turned back again, by suitable processes, to the ordinary form, also without gain or loss of weight. Phosphorus is not the only elementary substance that is capable of this kind of change. Indeed the general term *allotropism* has been applied to the tendency of elementary substances to undergo internal changes, by reason of which their chemical properties are temporarily modified without gain or loss of weight, and therefore independently of chemical combination or decomposition.

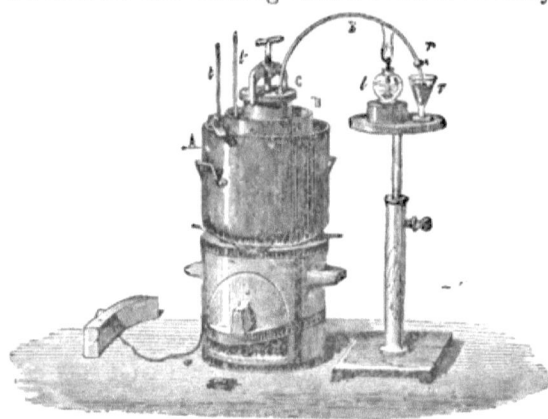

Fig. 58.—Coignet's apparatus for production of red phosphorus. Ordinary phosphorus is placed in a cast-iron vessel c; it is then heated ten or twelve days, an even temperature being maintained by the two iron jackets, one enclosing sand, the other holding fusible alloy.

Chemical Properties of Phosphorus.

The chemical properties of phosphorus are wide in their range; that is, it combines with many of the chemical elements. Thus it unites with hydrogen in more than one proportion, and thereby forms several compounds. As might be expected, they are all exceedingly combustible; one of them in particular, called phosphuretted hydrogen, takes fire at ordinary temperatures immediately upon coming in contact with the atmosphere. Its production affords opportunity for a beautiful experiment, though a somewhat dangerous one. When the gas is produced in a retort, it may be made to bubble through water in the form of vapor in company with various gases generated at the same time. Then, as it reaches the surface, it instantly takes fire, the phosphorus burning to a white, smoke-like substance which usually floats away in forms similar to those of smoker's rings. The smoke consists of minute particles of a solid, called phosphorus pentoxide, and expressed by the formula P_2O_5. This is evidently the product of the combustion

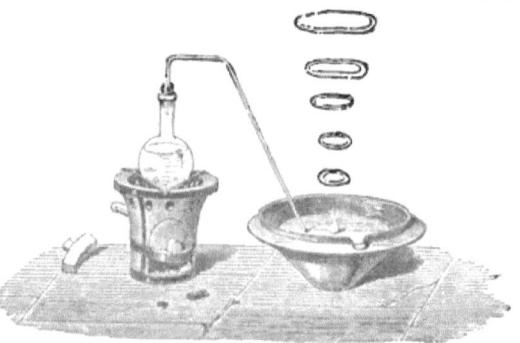

FIG. 59.—Phosphuretted hydrogen gas, of the spontaneously inflammable variety, taking fire in air and forming smoke-rings.

of that phosphorus which is a part of the inflammable gas. The shape of the rings is due to a mere mechanical circumstance and the same in effect as that afforded by the lips of the smoker while producing rings. Indeed if a paper box, having a round hole on one side, be filled with smoke of any kind, sharp blows upon the opposite side will drive out portions of the smoke in such a way as to produce similar rings. Such rings are often seen on a still day puffed out of the smokestack of a locomotive, and they are sometimes produced by the discharge of a cannon in still air. The fact is that in all these cases the portion of smoke producing a ring advances through the opening with a sudden impulse, the edge of the opening retarding those particles that pass nearest to it. Thus the delayed particles acquire a tendency backward and inward which starts them on the peculiar series of circular courses, which in the grand aggregate give rise to the rings.

As has more than once been stated, phosphorus has a marked affinity for oxygen. It burns in any vessel containing air, combining with oxygen in such a way as to readily deprive the air of the entire amount of this element contained in it.

The chemical change is represented by the following equation:

$$\underbrace{\underset{\substack{\text{One molecule of} \\ \text{Phosphorus,} \\ 124 \\ \text{parts by weight.}}}{P} + \underset{\substack{\text{Five molecules of} \\ \text{Oxygen,} \\ 160 \\ \text{parts by weight.}}}{5O_2}}_{284} = \underbrace{\underset{\substack{\text{Two molecules of} \\ \text{Phosphorus pentoxide,} \\ 284 \\ \text{parts by weight.}}}{2P_2O_5}}_{284}$$

When the operation is performed in a tall jar, the oxide of phosphorus produced falls as abundant flakes having a snow-like consistency. When these flakes are thrown upon water they chemically combine with the water, affording much heat and producing a hissing sound which is the evidence of it. The liquid now acquires a sour taste referable to the fact that phosphoric acid has been produced.

The chemical change is represented by the following equation:

$$\underbrace{\underset{\substack{\text{One molecule of} \\ \text{Phosphorus pentoxide,} \\ 142 \\ \text{parts by weight.}}}{P_2O_5} + \underset{\substack{\text{Three molecules of} \\ \text{Water,} \\ 54 \\ \text{parts by weight.}}}{3H_2O}}_{196} = \underbrace{\underset{\substack{\text{Two molecules of} \\ \text{Phosphoric acid,} \\ 196 \\ \text{parts by weight.}}}{2H_3PO_4}}_{196}$$

Phosphoric acid is the starting point of an immense series of salts called phosphates. One of these, calcic phosphate, we have already referred to as existing in bones and in apatite.

Friction Matches.

The earliest method of producing flame appears to have been by the friction of pieces of dry wood in contact with dry leaves or similarly combustible substances. This method travelers have found to be still in use among tribes of a low stage of development. The next method seems to have been by the use of flint and steel and tinder. When the flint is sharply struck against the steel, it tears off minute particles of the metal, and these fragments are heated to the luminous point by the violence of the stroke; if they are made to fall upon the tinder, this easily combustible material takes fire; from its burning, a candle or lamp may be lighted. But the flint and steel and tinder must be dry and in good order to produce the best results; even then considerable skill is demanded. So it is easy to see that mankind has often preferred to *preserve a flame once lighted*, and then communicate this to another and another from time to time, rather than to go to the trouble of exciting a new combustion when fire was needed. And it is easy to appreciate the usefulness to its possessor of a flame once kindled—and the serious inconvenience resulting from its extinction. Thus we can readily comprehend how nations have adopted fire as a sacred agent, to be preserved continuously unextinguished, and to be guarded with religious care.

The flint and steel method has ample illustration as to its principle, not only in familiar cases like sparks from the horse's hoof, but also in many processes in factories and machine shops. Here it is well known that the grindstones used for finishing articles of iron and steel send off from their work an uninterrupted current of minute chips of the hot and luminous metal.

The flint and steel method of obtaining fire held its own until about sixty years ago. In 1829 a kind of chemical match was devised, and soon after, in 1832, a true friction match containing phosphorus was brought into use. The *principles* upon which the phosphorus match depend are but very slightly different from those involved in the use of the flint and steel. Thus in the friction match the rubbing upon the rough surface is a mechanical process which generates heat, just as any blow or any friction does. In the case in question the amount of heat is small, but it is sufficient to set on fire the small amount of phosphorus on the tip of the match; the phosphorus sets on fire the sulphur which coats over the end of the match; the sulphur in burning sets on fire the wood of the match, and here the combustion has reached a stage at which it is easily communicated to larger masses of material. In the finer kinds of wooden matches, in order to avoid the objectionable smell of the burning sulphur, this latter substance is sometimes replaced by a thin coating of wax upon the end of the stick. In this case, other chemicals are added to the tip of the match, in order to make the combustion more active.

Friction matches of the ordinary kind are now so abundant and familiar everywhere that the exceeding usefulness, convenience and importance of the match as a device or invention, is apt to be overlooked. It is not intended to dwell here upon this subject, however, for perhaps what has been said of the appliances for lighting used in the past, renders unnecessary further presentation of the principles utilized in the little tapers of to-day.

As an article of manufacture, the individual match is so small that it is not easy at first to appreciate the greatness of the commercial interest it represents. Thus it is estimated that in Europe alone fifty thousand persons are constantly employed in the manufacture of the various kinds of matches. Again, though the amount of phosphorus used in each match is very minute, its sum-total is no less than a thousand tons a year. The value of the annual product of this industry is not far from fifty millions of dollars.

If there were introduced here an account describing at length the manufacture of the friction match—commencing at the beginning with the special kind of wood employed and the processes used for its subdivision into the requisite fragments, continuing even so as to explain the various contrivances for packing the finished product — that description might be of interest; but the special topic seems to be more properly the preparation and application of the material at the tip of the match. The sticks having been prepared, they are placed, by machine, in frames capable of containing large numbers of them. They are first sulphured, that is their ends are dipped in melted sulphur and it is allowed to harden upon them. For the finer grade of matches however, the sulphur must be dispensed with, and instead the sticks are dipped into melted wax.

In any case, they are next tipped with the highly inflammable material, this process being called chemicking. The inflammable paste is prepared in large quantities by mixing the proper ingredients in a kettle surrounded by boiling water. First, a solution of an appropriate gum or glue is made. When it has attained a proper consistency, the phosphorus is introduced little by little. The whole mass is then slowly but thoroughly agitated with a wooden stirrer until the phosphrous is diffused through the mass. Finally, other ingredients, such as potassic nitrate or binoxide of lead or manganese dioxide, which favor combustion, are added; and certain coloring matters, such as Prussian blue or vermilion, are introduced. Here is a German recipe for making this paste:

Gum,	16 parts.
Phosphorus,	9 parts.
Potassic nitrate,	14 parts.
Manganese dioxide,	16 parts.

As has already been intimated, all of these substances, except the phosphorus, may be replaced by others, according to the style of the article to be manufactured or the views of the maker. The process of chemicking consists in dipping the sulphured ends into the inflammable paste, which for this purpose is spread out on a stone slab. Finally, the tips are coated over with a thin varnish to protect them from absorption of moisture.

At present the manufacture of friction matches is carried on to a very large extent in Sweden, and that country, it is now stated, produces about seventy-five per cent of all the matches made in the world. In Sweden, too, are largely manufactured what are called safety matches. The safety matches are tipped with a composition of potassic chlorate, potassic dichromate, red oxide of lead, and sulphide of antimony. Under ordinary circumstances friction will not set these matches on fire. In lighting, they must be rubbed on a prepared surface which contains principally red phosphorus and sulphide of antimony. When the match is rubbed upon this surface, the potassic chlorate of the match and the red phosphorus of the friction-surface start a chemical combination which extends to the other

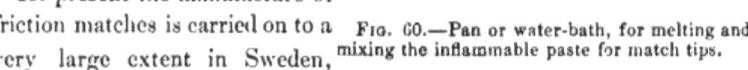
Fig. 60.—Pan or water-bath, for melting and mixing the inflammable paste for match tips.

materials on the tip of the match. Safety matches, then, involve an invention which in accomplishing its purpose, affords a twofold advantage. In the first place, as the match lights only on the prepared surface, the danger of conflagrations from accidental ignition of them is very largely reduced. This costly feature of the ordinary phosphorus match would be largely, if not entirely, done away with by the general use of the safety match. In the second place, the use of *red phosphorus* has the advantage of saving human lives in other ways. Thus it spares the operatives, employed in this business, the liability to the phosphorus disease already mentioned. Again, ordinary phosphorus is very poisonous; in fact the tips of matches containing this substance have not only often produced the death of children who have tasted them, but such matches have often been used in cases of intentional suicide. Of course as safety matches contain no phosphorus, these forms of poisoning cannot arise from them.

A flame of fire, as a visible and tangible thing, has in all ages been accepted as a symbol which appropriately typifies enlightenment of the mind and soul. This favorite and beautiful figure loses none of its fitness when narrowed in its application to the aspects of these subjects in their peculiarly modern forms. For in the friction match, whose cheapness brings it to the hand of every human being however low his degree, we may discover the type of that opportunity for enlightenment offered to individuals whose circumstances seem most humble and even forbidding. The one is the invention of modern science; the other the gift of modern laws, of modern theories of the rights of men, of modern schools, libraries, and newspapers, of the modern printing press, telegraph, and railroad.

READING REFERENCE.

Friction Matches.

Schrötter, A. v.—Chem. News. xxxvi, 207, 219, 259.

CHAPTER XXIII.

CARBON.

ARBON exists in nature in a multitude of forms. It is rarely found in the pure and uncombined condition, though certain well-known substances possess it in large quantity.

Ordinary Charcoal.

Probably the most familiar and representative form of carbon is that known as charcoal. But charcoal is rarely free from other chemical elements, and a distinction ought to be made between it and the absolutely pure form of the element under consideration. Charcoal is produced by the partial decomposition, under the influence of heat, of vegetable or animal substances. Thus charcoal is commonly prepared by piling wood into a conical heap, then covering it with earth and sods, and finally setting it on fire within. Certain portions of the wood are thus burned, while others are only charred. The wood is decomposed

FIG. 61.—Charcoal pit.

Fig. 62.—Tree trunks discovered in coal mines.

by the heat to which it is subjected; volatile materials generated by this decomposition are expelled, while there is left behind a solid matter consisting mainly of carbon, and called charcoal.

Animal Charcoal.

The same general treatment of certain animal matters, such as waste leather, gives rise to a finer kind of carbon called animal charcoal.

Fig. 63.—Charcoal burners at work.

Again, when bones are partly burned, they produce what is called bone-coal. The *mineral matter* of the bone undergoes no change by the heat; but the gelatinous matters which permeate it are decomposed, and they leave behind them the carbon deposited upon this mineral matter.

Lamp-Black.

Another material, closely assimilated to those already spoken of, is lamp-black. This is a product of the imperfect combustion of substances like oil, tar, resin, and the like, which are very rich in carbon. The tar or resin being set on fire is allowed to burn, but in an imperfect way, and so as

Fig. 64.—Manufacture of lamp-black.

to evolve a dense black smoke. The smoke flows into a chamber prepared for it, where the sooty material collects on the floor and walls. It is afterwards scraped up and put into packages for commercial distribution. In the English method of manufacture of lamp-black, the smoke is made to pass through a series of heavy canvas bags. From openings at the bottoms of the bags, the soot is afterward drawn out for packing.

Coal.

Anthracite coal and bituminous coal are both well-known compounds of carbon. Anthracite seems to be derived from bituminous coal which has been subjected in the earth to heat and pressure under conditions favorable to the expulsion of some of the more volatile constituents of the original bituminous coal. Both of these combustibles, when carefully studied, show distinct evidences of their vegetable origin. Plainly they are accumulated masses of the remains of a rank vegetation which flourished in an earlier period in the geological history of our globe. Careful observations made in the mines have revealed in the coal the existence of trunks of trees, branches, leaves, fruits, in various conditions from the one extreme of comparatively perfect preservation, to the other extreme in which the mineral preserves a mere impression of the original vegetable matter. These remains have made it possible to construct a complete botany of this period of geological history; and with but a moderate aid of the imagination, artists have been able to produce ideal landscapes representing these early forms of vegetable life as they flourished in the ancient ages.

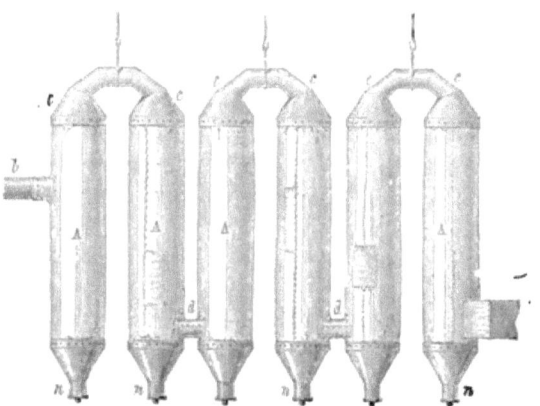

FIG. 65 — Bags in which lamp-black is collected in the English process of manufacture.

Graphite.

Closely allied to anthracite coal is that valuable material called graphite. This a very compact and comparatively pure form of carbon. It is familiarly known to every one in the black material used in lead pencils.

FIG. 66.—Imaginary landscape during the carboniferous era.

(202)

Graphite is commonly called black lead, though it is a well established fact that it contains no lead at all. Strangely enough graphite is remarkably incombustible under all ordinary circumstances. It is also—like other forms of carbon—infusible at the highest temperatures known. On account of these properties graphite finds use, though it must be deemed a somewhat anomalous one, in the manufacture of crucibles. When the precious metals are fused in such a crucible, at a high temperature in a glowing furnace, an interesting paradox is furnished. It is this: the coal—freely burning in the fire, and so furnishing the intense heat desired—is fundamentally of precisely the same chemical nature as the graphite of the crucible, which resists the heat and the combustion, and, while allowing the metals to melt, preserves them.

The Diamond.

The diamond is nearly pure carbon, crystallized. Perhaps it is not too much to say that it is the most striking and wonderful of all the forms of this interesting element. The costliness of the diamond is referable largely to its great rarity; for it is found in comparatively few portions of the earth.

The ancient Greeks and Romans highly prized the rare and precious crystal, which they obtained from India, and it was worn by them not only because of its costliness and beauty, but also because they believed that it served as a potent charm against alarms and enchantments; more important yet, they ascribed to it the power of preserving the peace and harmony of the family circle. Upon this point a French writer has wittily said: " Cette dernière vertu, je crois qu'il la possède encore quand le mari est assez riche pour acheter le bijou que sa femme ambitionne de porter!"

The East Indies, the Cape of Good Hope and the Brazils may be said to be the principal sources of this gem. In Brazil the search for diamonds is systematically conducted. The diamond bearing soils are carefully pulverized in vessels of water, under the direction of experienced inspectors. The work is done by slaves who prosecute their search under the stimulus of the well understood rule that he who finds a diamond weighing seventeen and one-half carats or more, publicly receives his freedom as a reward. Notwithstanding the systematic labor applied to the search for these gems

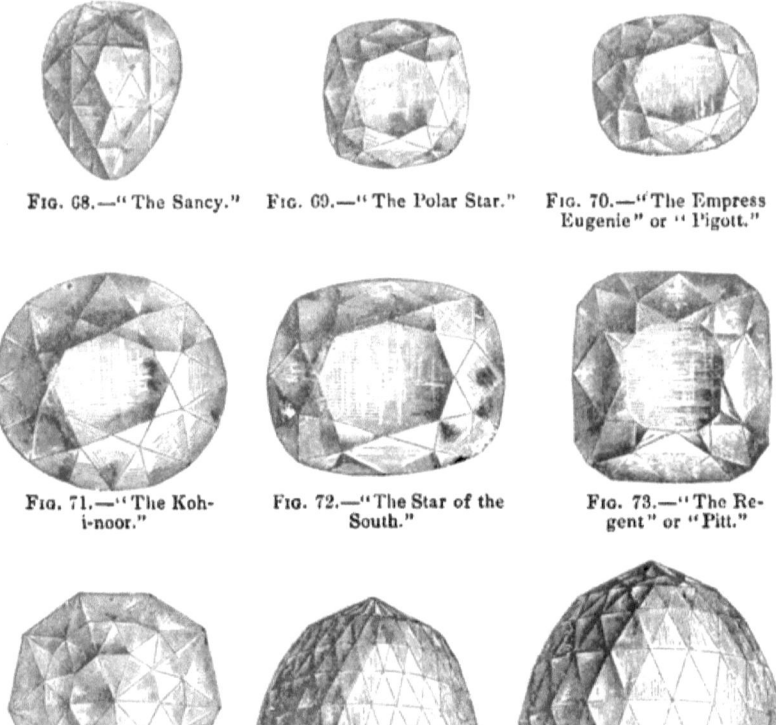

FIG. 68.—"The Sancy." FIG. 69.—"The Polar Star." FIG. 70.—"The Empress Eugenie" or "Pigott."

FIG. 71.—"The Koh-i-noor." FIG. 72.—"The Star of the South." FIG. 73.—"The Regent" or "Pitt."

FIG. 74.—"The Grand Duke of Tuscany." FIG. 75.—"The Orloff." FIG. 76.—"The Grand Mogul."

The great diamonds of the world (natural size).

and the fascination naturally attending undertakings of this sort, the wealth of Brazil is derived to a vastly greater extent from its agricultural products than from its mines. Thus it is stated that from 1740 to 1822, a period of more than eighty years, the diamond mines yielded but little more than

FIG. 77.—Transportation of diamonds under military protection.

$17,000,000. On the other hand the value of coffee exported in a single year has sometimes been double or even more than double this amount. Thus

FIG. 79.—Diamond cutter at work.

in the year 1859 the coffee exported was valued at above $28,000,000; and in 1873 the quantity of this article exported was valued at above $60,000,000.

The larger gems are exceedingly rare. On this account the money value of diamonds increases in a far more rapid ratio than the weight.

The Cutting of Diamonds.

The cutting of diamonds as an art has been known for but a few centuries, and the perfection with which it is at present conducted is of much more recent date. Of course the process is an extremely delicate and

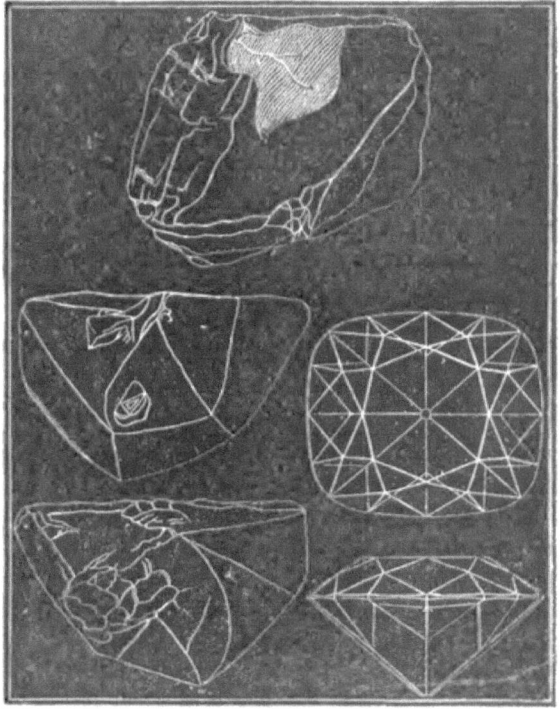

Fig. 80.—Diamond known as "The Star of the South"; before and after cutting.

important one because it involves splitting off portions of the gem so as to reduce it to the exact geometrical shape previously decided upon. That form called the *brilliant* is the one commonest produced at the present day. The business of cutting diamonds has been for a long time concentrated in the city of Amsterdam in Holland. Here, among a Jewish population of twenty-eight thousand persons, ten thousand are employed exclusively in working on diamonds. In many cases diamonds are subject to a very large relative loss of weight by the process of cutting. Thus the Koh-i-noor, when brought from India as a gift from the East India Com-

Fig. 81.—Diamond polisher at work.

pany to the English Crown, was in the rough state and weighed one hundred and eighty-six carats. It was afterwards cut in Amsterdam, by which process it suffered a loss of weight variously stated as from eighty to one hundred carats. After the first trimming, the gem is carefully polished by rubbing it gently against a revolving plate upon which is a mixture of oil and diamond dust.

Up to the close of the last century the nature and composition of the diamond had been a subject of interesting discussion among students of

Fig. 67.—Diamond washing in Brazil.

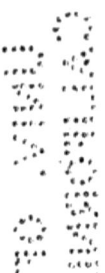

natural science. At the period mentioned however, the question was settled by Lavoisier and other scientific investigators, who clearly proved

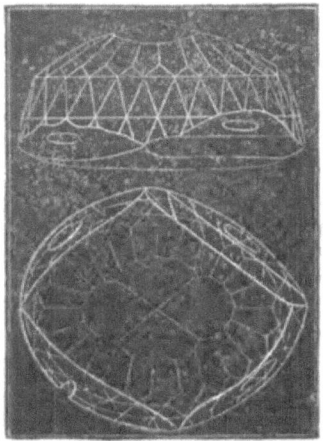

Fig. 82.—The Koh-i-noor before cutting. Fig. 83.—The Koh-i-noor after cutting.

that the diamond underwent complete combustion in oxygen and that as a result carbon dioxide gas was generated.

Other Natural Forms of Carbon.

In addition to the well-known forms of matter containing carbon, and already described, there are yet many others.

Thus it is found in the atmosphere, as has already been explained, in the form of carbon dioxide. This gas exists in the air in small relative proportion, but in enormous aggregate amount.

As a natural carbonaceous substance petroleum, too, ought not to be forgotten. This wonderful and useful substance stored up beneath the surface of the earth, in incredibly large quantities, owes its chief value to its wealth of carbon. It is composed of carbon and hydrogen, but the former is the constituent to which is referable the beautiful light it affords.

Again, the marble and the limestones of the globe contain enormous quantities of carbon. These minerals consist principally of calcic carbonate ($Ca\ CO_3$); and the carbon makes up about one-eight of this

Fig. 84.—View in the vicinity of the Pennsylvania oil wells.

substance. Since some whole mountain chains consist chiefly of limestone or marble, it is plain that the total amount of carbon in these forms must be very large.

With few exceptions, all animal and vegetable matters contain carbon —a substance which appears to perform its most important offices in connection with the kingdoms of life. Indeed it has been called the characteristic element of animal and vegetable compounds. So vast is the variety of these compounds already recognized that it is hardly conceivable that man can ever be able to acquire an acquaintance with all those as yet undetected.

Infusibility of Carbon.

Carbon differs from most solid substances in the fact that it is infusible at the highest temperatures to which it has yet been subjected. And since in the elementary form it has not been changed to the liquid state, much less has it been brought to the *gaseous* condition. Indeed this stability and fixedness of carbon is one of its most valuable attributes. Thus this characteristic is a principal one that renders it specially appropriate for use in the pencils employed in electric lighting. It is true these pencils slowly burn away. But some combustion ought to be expected when it is remembered that the electric current, flowing from one pencil to the other, affords an intense heat as well as brilliant light. But it is a general law that substances give out the most intensely brilliant white light when they neither

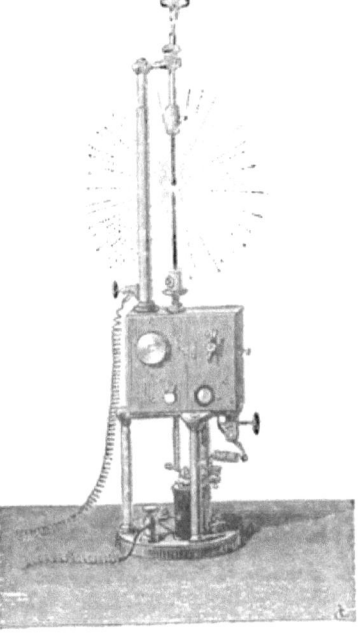

FIG. 85.—Automatic regulator whereby the carbon pencils of the electric light are maintained at the proper distance apart.

liquefy nor volatilize, and to this principle—exemplified in the carbon pencils—must be referred the brilliancy of the electric light.

Decolorizing Power of Carbon.

Carbon, whether in the form of wood charcoal, animal charcoal, or bone-coal, has a wonderful power of decolorizing liquids. Even more compact carbonaceous matter, such as anthracite coal, possesses this same property though, as might be expected, to a much inferior degree. Thus if a colored solution is strained through a considerable quantity of one of these forms of carbon, the latter substance absorbs the coloring matter and the liquid passes through practically colorless. On account of this wonderful power bone-coal is used in the arts in enormous quantity in many processes where liquids must be decolorized. The sugar refining industry affords a prominent example upon this point. Here, enormous quantities of bone-coal are used for the purpose of whitening the syrups before crystallizing the sugar.

Fig. 86.—Magnified view of the carbon terminals used for the production of the electric light.

FIG. 87.—Excavations carried on at night by aid of the electric light.

Charcoal has also a similar, and yet more striking, property of absorbing offensive gases. Thus, tainted meat packed in freshly burned charcoal quickly loses its odor—which is absorbed by the coal—and the meat then becomes sweet and wholesome.

Chemical Properties of Carbon.

The chemical properties of carbon are by no means less wonderful than the characteristics already referred to. It is very inert at low temperatures; but at high temperatures, it manifests chemical activities of extraordinary vigor. Thus at high temperatures carbon withdraws oxygen from almost any other elements known, in this way manifesting chemical force superior to that possessed by any of them.

Fig. 88.—Colored liquid filtered through charcoal, and thereby decolorized.

The Great Number of Compounds Formed by Carbon.

The vast number of the compounds of carbon seems to be referable to two fundamental properties with which it is endowed by nature. One of these is the fact that the atom of carbon possesses four points of attraction. This matter need not be explained here as it has already been discussed at sufficient length. But the fact may be conveniently represented to the eye by a symbol like the following:

$$-\overset{|}{\underset{|}{C}}-$$

The other property referred to is this: carbon—unlike most other elements—has a peculiar capacity by virtue of which atoms of it may join together in either short or long chains, and afterwards may gather other elements to the various parts of the chain. One among the many ways in which carbon takes part in forming such compounds may be represented by the simple diagram shown in the margin:

The number of compounds of this character already known to chemists is very large; it suggests the probability that there is no distinct limit to the number of atoms that may be linked in a continuous chain in this way. Moreover a given chain may have attached upon its sides or ends one or more additional chains of elements or compounds and thus give rise to an all but infinite number of substances with almost infinitely varied properties.

READING REFERENCES.

Coal, and the Coal-mines of Pennsylvania.
    Harper's Magazine. xv, 451.

Diamonds.
    Scribner's Monthly. v, 529.
    Harper's Magazine. xix, 466; xxxii, 343.

Diamond Fields of South Africa.
    Harper's Magazine. xlvi, 321.

CHAPTER XXIV.

COMPOUNDS OF CARBON AND OXYGEN.

HILE compounds of carbon and *hydrogen* are very numerous, those already known being numbered by hundreds, the affinities of *oxygen* and carbon give rise to a strikingly different result.

When combined with oxygen alone, carbon forms but two compounds. These are expressed by the following names and formulas:

Carbon monoxide, CO.
Carbon dioxide, CO_2.

Carbon Monoxide (CO).

This gas is most familiarly known as that one which often plays upon the surface of a hard coal fire and burns there with a dark blue, feebly luminous, flame. Most of the phenomena of its production and final burning may be presented as follows: When an ordinary coal fire, burning in a stove, is amply supplied with air at the bottom, the oxygen of the air burns the lower portions of carbon into carbon dioxide. Next, this carbon dioxide is carried up, by the draft, between any masses of fresh coal that may be upon the top of the fire. This fresh coal has itself affinity for oxygen under the circumstances just described as prevailing. As a result, each molecule of carbon dioxide from the lower portion of the fire yields one of its atoms of oxygen to an atom of carbon in the upper part.

The chemical change is represented by the following equation:

$$CO_2 \quad + \quad C \quad = \quad 2CO.$$

One molecule of	One atom of	Two molecules of
Carbon dioxide,	Carbon,	Carbon monoxide,
44	12	56
parts by weight.	parts by weight.	parts by weight.

$$56 \qquad\qquad\qquad 56$$

As a result, therefore, carbon monoxide is formed and escapes as a colorless gas from the top of the fuel; there, if the upper door of the stove admits a sufficient amount of air, the carbon monoxide combines with the oxygen of this air, and burns with the blue flame already referred to, and so produces carbon dioxide again.

This chemical change is represented by the following equation:

$$2CO \quad + \quad O_2 \quad = \quad 2CO_2$$

Two molecules of	One molecule of	Two molecules of
Carbon monoxide,	Oxygen,	Carbon dioxide,
56	32	88
parts by weight.	parts by weight.	parts by weight.

$$88 \qquad\qquad\qquad 88$$

The carbon monoxide is a very poisonous gas, far more injurious to health than carbon dioxide.

Carbon Dioxide (CO_2).

This substance and its manner of production have been referred to more than once in preceding chapters. A more extended notice of it, however, is appropriate to this place.

It has already been stated that carbon dioxide exists ready-formed in nature—notably in the atmospheric air. Its principal natural source in the atmosphere is the combustion of fuel; for almost all fuel is carbonaceous. Thus coal, wood, oil, illuminating gases, are all highly carbonaceous substances, and one of the principal products of their combustion is the gas now under consideration.

As has already been described, the respiration of animals is closely connected with a real combustion in the living being. It is true that this

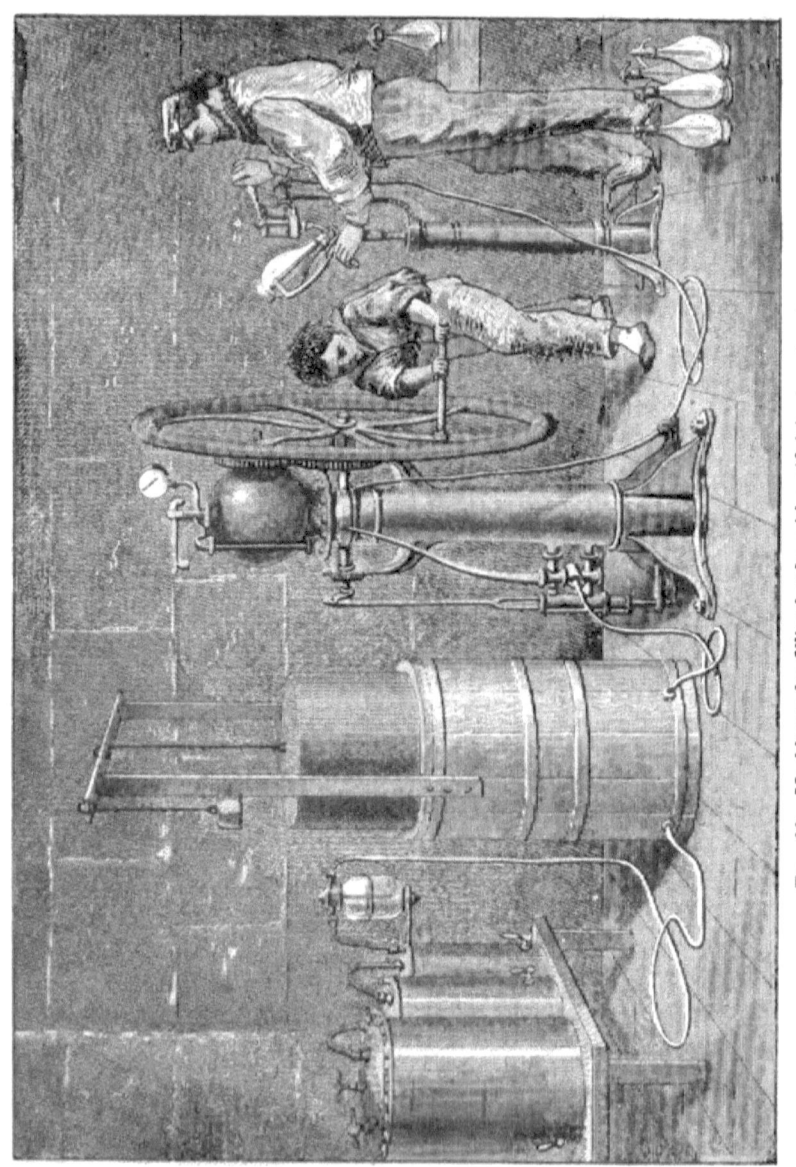

FIG. 90.—Machinery for filling bottles with artificial mineral waters.

sort of combustion is not attended by the evolution of light; it is productive of heat, nevertheless, and the heat afforded by respiration is an important factor in the sustenance of animal existence. For this heat not only enables the living being to endure the chilling effects of the winter's cold; it also keeps the temperature of the internal organs up to that point which is necessary for the proper performance of certain animal functions—of which digestion is a most important example. Now by this combustion carbon dioxide is generated just as truly as would be the case if the flesh of the living animal were consumed in a glowing fire.

Fig. 89.—Production of carbon dioxide by combustion of a diamond in oxygen gas.

The product of respiratory combustion is the same carbon dioxide as that recognized in well-established burnings. The quantities of carbon dioxide evolved by man and certain of the domestic animals, in each hour of their existence, have been calculated. They are approximately stated in the following table:

A man exhales 4 gallons carbon dioxide per hour.
A dog " 4½ " " " " "
A horse " 50 " " " " "
An ox " 70 " " " " "

And M. Boussingault has calculated that the approximate amount of carbon dioxide produced in the city of Paris during a single twenty-four hours is as follows:

Amount produced by living animals, . 55,000,000 cubic feet.
Amount produced by burning of various
  kinds of fuel, ... 27,000,000 cubic feet.

Total CO_2 produced in twenty-four hours, 82,000,000 cubic feet.

There are certain other natural sources of carbon dioxide that are worthy of passing mention. Thus in many parts of the world the gas is continually evolved not only from active volcanoes but also from extinct ones. Again, another interesting source—though not in the aggregate a very important one—is found in natural mineral springs. In these the water often comes to the surface highly charged with carbon dioxide, and the gas, escaping into the air, imparts to the water its well-known bubbling appearance.

Experiments with Carbon Dioxide.

For chemical purposes carbon dioxide is commonly produced by the action of an acid upon some one of the salts known as carbonates. Accordingly chlorohydric acid and calcic carbonate (that is, common marble) when brought together produce carbon dioxide. This fact may be readily shown by the performance of a simple but interesting experiment. The operation may also serve for the display of some of the principal properties of the gas.

The experiment in question may be conducted advantageously somewhat as follows: Provide two convenient glass jars—such as candy jars or preserve jars; also a short candle, a piece of copper wire, a bottle of chlorohydric acid and some fragments of white marble. Now attach the candle to the wire and after lighting the former let it down into the jars, still burning. The combustion continues because the jars are full of air and contain ample quantities of oxygen. Next withdraw the candle and extinguish it for a moment. Now place in the bottom of the larger jar some chlorohydric acid and into it gently drop some of the fragments of marble. Effervescence immediately commences. A careful examination of effervescence shows that in this, as in other cases, the process consists in the evolution of a gas from a liquid. In the case in question a colorless gas is plainly evolved, and this gas is carbon dioxide.

The chemical change is represented by the following equation:

$CaCO_3$	+	$2HCl$	=	CO_2	+	$CaCl_2$	+	H_2O
One molecule of Calcic carbonate,		Two molecules of Chlorohydric acid,		One molecule of Carbon dioxide,		One molecule of Calcic chloride,		One molecule of Water,
100		73		44		111		18
parts by weight.		parts by weight.		parts by weight.		parts by weight.		parts by weight.

$$173 \qquad\qquad 173$$

After allowing the effervescence to continue for five or ten minutes, relight the candle and again lower it into the jar now containing carbon dioxide. If a sufficient quantity of the gas is present, the light will be promptly extinguished when the wick passes below the surface of the gas. The experiment displays at this stage the additional fact that the carbon dioxide is heavy, and in filling the jar it does so from the bottom upward. Now relight the candle and immerse it in the second jar; this is proved to contain air by the fact that the candle continues to burn. While it is still quietly burning there, pour gently upon it the carbon dioxide accumulated in the other jar. If the amount of this gas is large enough, it will fill the jar containing the lighted candle and so will readily extinguish the latter.

These experiments demonstrate simply and clearly, certain of the most important properties of carbon dioxide. Its action in extinguishing flame is to quench it, very much as water would. When the candle dips beneath the surface of the carbon dioxide, the flame expires simply from lack of that oxygen of the air which ordinarily supports the combustion. And this leads very naturally to the additional statement that, in similar fashion, living beings are drowned if immersed in carbon dioxide. For just as water prevents the access of air to the lungs, and then drowning ensues, so when the animal is beneath the surface of carbon dioxide he dies from the similar deprivation of air.

Effervescing Beverages.

A large quantity of carbon dioxide taken into the *lungs* is promptly fatal to animal life, and even a small increase of that gas, in the atmospheric air breathed, also produces a marked lowering of the vitality. It is an interesting fact however that when this gas is taken into the *stomach*, especially in its solution in water, it has a wholesome and stimulating effect.

When carbon dioxide is dissolved in water it seems to produce a true acid, though an unstable one. In accordance with the present nomenclature, this acid is called carbonic acid and is represented by the formula H_2CO_3. This substance is present as the main constituent, or as a subordinate one, in certain natural mineral waters, and in many simple effervescent beverages. Thus plain soda-water is merely a solution of carbon dioxide in water. Such solutions are now manufactured on a large scale, and by mechanical appliances are filled into siphonlike bottles in such a way that small quantities of the liquid may be withdrawn without loss of the principal stock of gas.

CHAPTER XXV.

ILLUMINATING GAS.

IT has been stated more than once that the compounds of carbon are very numerous; it might properly be added that their usefulness is no less striking than their number. This portion of the subject is so vast, however, that it is often discussed as an entirely separate branch, called the chemistry of the carbon compounds, and often organic chemistry. By organic chemistry is meant the chemistry of organic substances; and by organic substances is meant materials derived from those existences that possess organs. Now animals and plants, and they alone, possess organs; whence organic chemistry is described as the chemistry of animal and vegetable bodies. On some accounts, it is better defined as the chemistry of the carbon compounds; for while it was formerly thought that animal and vegetable beings involved in their processes a chemistry peculiar to themselves, this notion has long since been dispelled, and it is now clearly perceived that animal and vegetable compounds are governed by the same chemical laws as others.

But not only is the number of these organic compounds very great; the variety and importance of animal and vegetable matters give them a high degree of interest. Thus they include animal and vegetable juices, extracts, gums, resins, essences, remedial agents, bitter principles, acids, oils, coloring matters; and of the members of each one of these classes the name is legion. Moreover, the more any portion of the subject is studied, the more it seems to reveal a continually increasing complexity.

Most of the substances of the classes referred to are either compounds of carbon and hydrogen, or of carbon, hydrogen and oxygen in varied proportions, or they are compounds containing these elements and yet a few others combined with them.

It is manifest, from what has been said, that in a book like the present it is impossible to give any considerable discussion of the vast field offered by the organic compounds of carbon. It seems better to choose for description some important manufacturing operation that involves these compounds and that is on other accounts specially instructive. Accordingly the manufacture of illuminating gas is selected for consideration here.

The Manufacture of Illuminating Gas.

The material on which this industry is based is bituminous coal. This substance is clearly a vegetable product, though it is derived from a vegetation which lived, flourished and decayed in a period of pre-historic antiquity.

The manufacture of illuminating gas, although one of the most important of the chemical industries of today, had its beginning but little before the opening of the present century. A Scotchman named William Murdoch is generally credited with the first introduction, into considerable use, of burning gas made from coal. In 1798 he gained the opportunity to introduce his method of illumination into the engine works of Boulton & Watt, located at Soho, near Birmingham. From that date the manufacture and use of illuminating gas from soft coal has extended and expanded until it has reached its present enormous development.

General Principles of the Process.

The general principle of the manufacture is exceedingly simple. But its commercial growth has been assisted by the invention and application of a multitude of delicate and ingenious appliances.

If any person will take a glass test-tube, place in it a few fragments of starch, and will then heat the starch strongly over a lamp flame, he will readily detect three important effects. The first is that a mass of smoky gas or vapor pours out of the mouth of the test-tube. The second is that an oily or tarry liquid condenses, on the inside of the tube, and runs down in streams. The third is that at the close of the operation a mass of carbon remains in the bottom of the tube where the starch was. Now the various substances, that have been referred to as produced by the heating process, are referable to the decomposition of the molecules of starch.

A Similar Operation on a Large Scale.

In the manufacture of illuminating gas on a large scale there are developed practically the same series of phenomena as those noted in the experiment with starch just referred to.

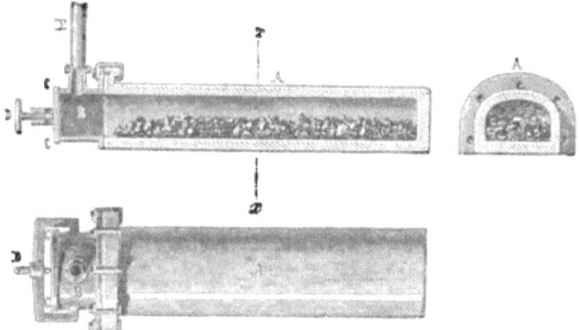

FIG. 92.—Three views of a gas retort.

In the manufacture of illuminating gas, instead of starch as just described, soft coal or bituminous coal is used.

In place of a lamp, a large row of furnaces is employed to supply the heat.

Instead of glass tubes, those of earthenware, ten or twelve feet in length and between one and two feet in diameter, are used. These tubes, called retorts, are placed in a horizontal position and so that the flame of the fire in the furnace may sweep around them and raise them to a cherry-red heat. At the front end of the retort is attached a door to prevent the escape of the gases generated, and there is also a suitable pipe to carry these gases forward to those other portions of the works which serve to perform upon the crude gas certain necessary purifications; these are: *First*. The condensation of condensable vapors. *Second*. The removal of objectionable gases.

The operations spoken of show that the gas must be carried from one portion of the establishment to another. Now illuminating gas is made up of material substances and although lighter than air yet they distinctly possess weight. Gas will not move of itself; to carry it from place to place the application of force by means of mechanical appliances is requisite. In fact it is discovered that what is called an *exhauster* is necessary for use in gas works. The exhauster is simply a kind of rotary pump which pulls the gas from the retorts in which it is first formed, and pushes it along through the various purifiers, to the gas holder in which it is stored. If the exhauster were not used, there would be a constant tendency to the creation of pressure in the retort, by virtue of which the gas would penetrate the earthenware into the fire, and so become a source of loss.

Fig. 93—Section showing five retorts in place.

From what has been said it will be easily comprehended that the essential parts of a gas-works are the following:

First. The furnace.

Second. The retorts, in which the coal is heated.

Third. The hydraulic main: a trough of water in which the gas is cooled, and which also serves as a gate, through which the gas can pass forward toward the purifiers but not backward toward the retort.

Fourth. The out-door condensers, in which the gas is cooled and some of its vapors condense to tarry liquids.

Fifth. The scrubber, in which the gas is cleansed by a spray of water.

Fig. 91.—View in gas-works; drawing coke from the retorts.

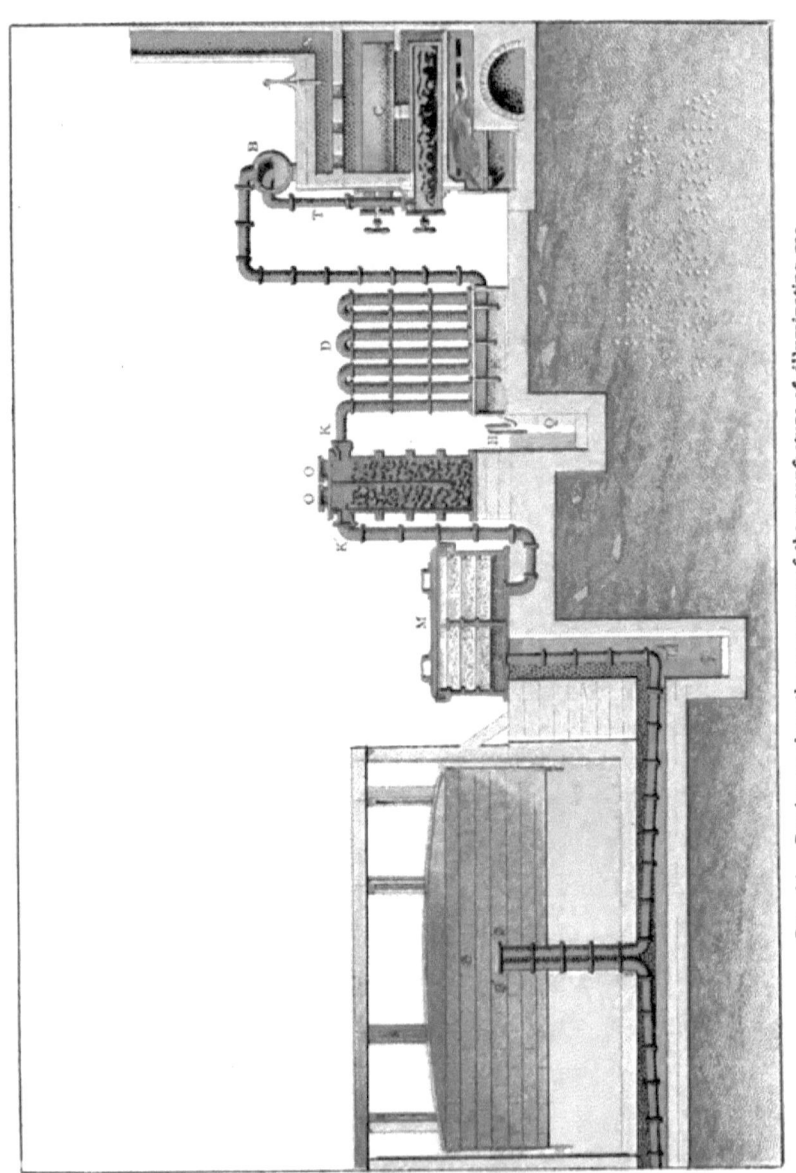

Fig. 94.—Section to show the processes of the manufacture of illuminating gas.

Sixth. The purifiers, where sulphuretted hydrogen, and some other objectionable gases are removed.

Seventh. The gas holder, in which the finished gas is collected and stored prior to delivery to consumers.

The processes by which these various appliances are used in the manufacture of illuminating gas may be briefly sketched as follows:

A suitable quantity of soft coal is placed in an even layer on the bottom of the *retort*. Gas at once forms and streams out of the open door. The door of the retort being now quickly closed by the workmen, the gas passes out through an exit pipe—called the dip-pipe because it dips into the water of the *hydraulic main*. The gas bubbles up from the dip-pipe through the water. Once delivered in the hydraulic main, the gas cannot go back to the retort.

Next, the gas passes through the *condensers*, a series of connected up-and-down pipes. As these condensers stand in the open air they cool the gas so that it deposits tarry liquids that, until this stage, have been suspended in it in the form of vapor.

Next the gas flows to a large iron box, called the scrubber. In different works the *scrubber* varies considerably in outward shape and internal arrangements. Its essential office however is to wash the gas, and it does so by the use of water which is applied to the gas either in sprays or thin films. Ammonia gas is the principal substance absorbed by the water in the scrubber. Indeed the liquor thus produced is the main commercial source, at the present day, of ammonia and its compounds.

The gas next goes to the *purifiers*. These are large iron boxes supplied with a multitude of shelves upon which, in most works, dry quicklime is spread. The quicklime absorbs sulphuretted hydrogen and some other acid gases. From these purifiers the gas is carried on to the *gas holder*.

The Distillation of Coal, Chemically Considered.

Under the influence of the high temperature of the gas furnace, the soft coal in the retorts undergoes decomposition. As has before been intimated, three distinct classes of substances are produced: Solids, which are left in the retorts; liquids, which are condensed in the various coolers; gases, which pass on the gas holder.

First. The solids. These are principally two kinds of carbon. One is coke,—the principal solid matter found in the retorts as a residue from the soft coal after the latter has ceased to evolve gas. It is merely a form of carbon, somewhat spongy in its structure. It is sold for use as fuel. Beside this the retorts accumulate a sort of scale of a very different form of carbon called gas carbon. It is extremely hard and almost non-combustible, being even very difficult to remove from the retorts. It is at present somewhat used in the manufacture of the carbon pencils employed in electric lights of the arc variety. Prior to this use it found scarcely any commercial outlet.

Second. The liquids. The first condensation of liquids takes place in the hydraulic main where tarry and oily matters condense and accummulate, and are drawn off from time to time into the tar well. Again in the condensers there is a still further deposition of liquids, also tarry and oily in their nature.

These liquids consist of very complicated mixtures of carbon compounds, but they are of the most interesting character. In the earlier stages of the manufacture of coal gas they were regarded as mere nuisances. Little by little however chemists have learned to separate the intermingled products, and have thus been able to obtain a number of substances of striking interest and usefulness in the arts. Among the multitudes of substances that go to make up the liquid called coal-tar, some are as yet hardly classified · others are distinctly recognized and have uses of great commercial importance. Of these latter, two will be mentioned here. These are anthracene and benzole.

The substance called *anthracene*, a compound of carbon and hydrogen ($C_{14} H_{10}$), has within the last ten years sprung into the highest commercial importance. This is referable to the fact that it has been found to be a suitable material from which, by chemical processes, there may be manufactured a substance known as alizarine, besides other equally valuable and interesting compounds. Alizarine was previously recognized as the coloring matter of chief value in madder root, a substance that has been used as a dye-stuff for above a thousand years. The alizarine, whether of madder or from anthracene, is a coloring matter of the highest value and usefulness. It affords turkey-red and other colors that are very important because they are extremely brilliant and extremely fast. Its *artificial* manufacture, from the anthracene of the filthy and offensive coal-tar, is one of the greatest triumphs of this or any age.

Another substance found in the coal-tar is *benzole*, a compound of carbon and hydrogen having the formula C_6H_6. This is the principal material from which, by a variety of well understood though complicated chemical processes, the well-known aniline colors have been produced. While these colors may well command the admiration of all, on account of their unsurpassed beauty and brilliancy, they are of especial interest to the scientist by reason of the chemical laws they illustrate. The preparation of these colors, as a group, ranks second only as a chemical achievement to that of artificial alizarine.

Third. The Gaseous Products. The gases generated in the process of the coal-gas manufacture are extremely numerous; some of them are of high illuminating power, of which that called ethylene (C_2H_4) is an excellent example. Again there are some that are combustible, but yet are of slight illuminating power. Substances of this class are present in the finished product. Hydrogen and carbon monoxide (CO) may serve as examples. There are always present also gases that are either injurious to the illuminating power or are otherwise objectionable. For example, nitrogen is always present, and it is not practicable to remove it from the gas. It contributes nothing to the value of the product. Again certain sulphur compounds, like sulphuretted hydrogen, are usually present. These indeed burn, but they give rise to offensive and unwholesome oxides of sulphur.

The sketch thus given, while it but imperfectly describes the wonderful industry in question, with its various well contrived and delicate appliances, serves however to give some idea of the importance of the operation, from a chemical point of view, and the mine of rich materials its carbon compounds offer to chemical students.

CLOSING CHAPTER.

CHAPTER XXVI.

SILICON.

SILICON may well be considered important on account of its *quantity* in the earth, if on no other. In an earlier chapter it has been shown that oxygen exists in our globe—including its atmosphere and its oceans—in an amount equal to about one-half of the weight of the whole. Now silicon exists in a quantity equal to about one-fourth of this entire weight. In the earth however, neither of these substances exists in the uncombined form. These facts seem to involve as a necessary consequence that they exist in the earth, to a large extent, combined with each other; indeed this is found to be the case. The principal earthy matter of our planet is the compound of silicon and oxygen, existing either alone in the form of sand, quartz crystal and similar minerals, or else in combination with other well-known abundant earth materials, such as oxides of calcium, magnesium and aluminum. It has already been stated that carbon is the characteristic element of animal and vegetable matters; so silicon is the characteristic element of mineral matters. Thus granite and similar archaic rocks contain approximately twenty-five per cent of silicon.

In nature, silicon performs its important office as a constituent of rock material, with a fitness that is referable largely to the high degree of stability possessed by most of its compounds. The permanence of the materials of the earth's surface under the influence of heat, and water, and frost, and similar agencies is an illustration of this principle.

Silicic oxide (SiO_2), occurs on our globe in many different forms of which diatomaceous earth and rock crystal may be mentioned.

Diatomaceous earth is a powdery material found in abundant deposits in many parts of the world. Its characteristic structure, when examined under the microscope, reveals its nature; then it is seen to be made up of the shells of minute vegetable organisms called diatoms. These assume a great many beautiful forms, and some of them are checkered all over with markings of such extreme fineness that they have been used as test objects for

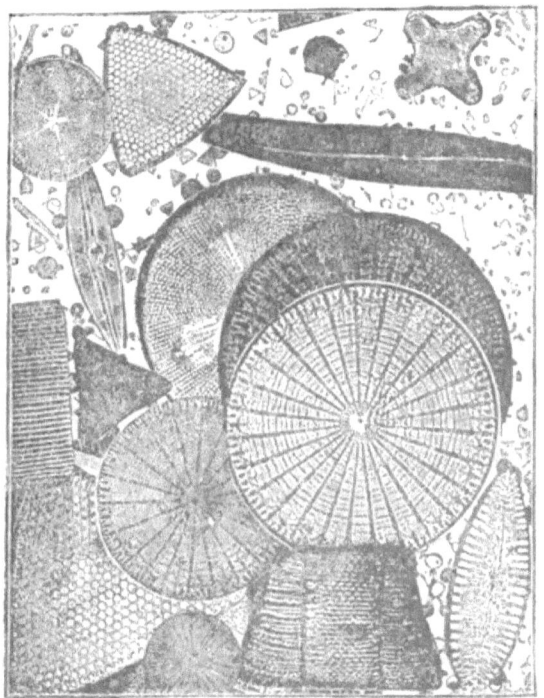

Fig. 95—Diatomaceous earth as seen through the microscope.

trying the resolving power of the objectives of microscopes. This kind of earth is employed, as has already been stated, in the preparation of dynamite.

Quartz sometimes occurs in colorless transparent masses of great beauty and clearness called rock crystal. The amethyst is the same substance slightly colored by compounds of the metal manganese, while quartz exists of a variety of other shades, in some of which it is prized as a gem.

Quartz generally assumes forms of a hexagonal tendency: they are often hexagonal prisms terminated by hexagonal pyramids.

Quartz and the finer and purer varieties of sand are used largely in the manufacture of glass. The silicic oxide here displays what may be expressed as its acid tendencies; for in the manufacture of glass it is fused

FIG. 96.—Mass of natural quartz crystals.

with sodic carbonate, and then the silicic oxide displaces the carbon dioxide from the sodic carbonate; as a result there is formed what must be regarded as a true salt, or a mixture of salts, that in the simplest kind of glass may be termed *sodic silicate*.

Closing Words. The course laid out in the preface is now terminated with silicon, as there planned. From scientific considerations, this is a natural ending; it seems to be appropriate on another account also. After the reader has been carried, in thought, among the various gaseous elements, that make up atmosphere and oceans, it seems suitable that we should say farewell to him upon the discussion of that element that may be called the characteristic material of our solid earth.

INDEX.

Abel, F. A. 182
Acid, Boric 157
—— Nitric 35, 168
—— Sulphuric 35, 37, 152
—— Chlorohydric 33, 190
—— Phosphoric 192
Affinity, Chemical .... 44, 46—52
Alizarine 226
Alkali Trade 94, 106
Allotropism 190
Amethyst 229
Ammonia 166, 174, 225
Anhydride 151
Anhydrite 146
Anthracene 226
Antimony, Sulphide of 148
Apatite 188, 189
Arsenic, Sulphide of 148
Atom 41, 42
Atomic Theory 52
Atmosphere 170

Bacon, Roger 178
Balard 100
Balloons 74
Barilla 107
Benzole 227
Berthollet 94
Berzelius 24
Beverages, effervescing 219
Binary Compounds 32
Biot 80
Black, Joseph 61, 63, 76
Bleaching Powder 93
Blende 146
Blowpipe, Compound 129
Bone-Ash 187
Borax 157
Boron 157
Boussingault 217
Brandt 187

Bromine 100
Carbon 197
—— Dioxide 173, 216
—— Monoxide 215
Cavendish, Henry 60
Charcoal 197
Chemistry, Scope of 7, 9
Chlorine 85
Cinnabar 146
Coal 201
Coal-tar 226
Coke 226
Compounds 32
—— Number of 8, 9
—— Organic 221
Copper, Sulphate of 51
Cotton 181
Courtois 108
Current, Galvanic 64

Daguerre 103
Dalton, John 52, 56
Davy, Humphry 60, 86, 108
De Rozier 78
Diamond 203
Diatoms 185, 228
Diffusion 69
Döbereiner 70
Dynamite 183, 185
Earth's Crust, Composition of 15
Elements, Chemical 9, 12, 14, 28
—— Names and Symbols of .. 19, 23
Equivalence 48, 72
Ethylene 227
Explosives 177

Fertilizers 188
Fluorine 112
Fireworks 180
Fulminates 181
Galena 146

INDEX.

Gas 63
—— Illuminating 221
Gas-Carbon 226
Gay-Lussac 80
Glaisher and Coxwell 81
Glass 229
Glycerine 184
Granite 228
Graphite 201
Gun-Cotton 181
Gunpowder 177
Haüy 15
Heat, unit of 71
Hydrogen 58

Iodine 105
Iron, Sulphide of 146

Janssen 83

Kelp 107
Lamp-Black 200
Larderel 160
Lavoisier 19, 21, 118
Lead 34
—— Sulphide of 146, 148
Leblanc Process 94
Liebig 101
Light, Calcium 131
—— Electric 211, 226
Limestone 209
Lithium 30

Marble 209
Mass 39
Matches 192
Mayow 121, 163
Metals 28, 29
—— Names of 16, 20
Mercury 30
—— Sulphide of 146
Moisture in the Air 173
Molecule 40
Montgolfier 74
Murdoch 222

Nobel 185
Nitrogen 162

Nitroglycerine 184
Nomenclature, Chemical .. 19, 20, 32, 33, 37
Non-metal 30

Oxide, Ferroso-ferric 66
Oxygen 117
—— Abundance of in the earth 14, 16

Palladium 70
Pelletier 89
Petroleum 209
Photography 93
Phosphorus 187
Phosphorus Necrosis 189
Phosphuretted Hydrogen 191
Potassic Chlorate 123
Priestley, Joseph 64, 118, 163
Puymaurin 115
Pyrites 146

Quartz 228

Roe 89
References, Reading, 10, 17, 24, 27, 31, 38,
  45, 56, 73, 84, 98, 111, 116, 133, 156, 161,
  186, 196, 214.
Rutherford 163

Scheele 85, 114, 118, 162, 187
Silicon 228
—— Abundance of 16
Sodium 67
Sugar 41
Sulphur 142
—— Dioxide 33, 149
—— Trioxide 33, 151
Sulphuretted Hydrogen 147

Ternary Compounds 34

Van Helmont 63
Varech 107
Vitriol, Oil of 152

Water 127, 134
Weight, Atomic 17

Zinc 68
—— Sulphide of 146, 149

www.ingramcontent.com/pod-product-compliance
Lightning Source LLC
Chambersburg PA
CBHW031351230426

43670CB00006B/501